Backward Fuzzy Rule Interpolation

Shangzhu Jin · Qiang Shen
Jun Peng

Backward Fuzzy Rule Interpolation

Shangzhu Jin
College of Electrical and Information Engineering
Chongqing University of Science and Technology
Chongqing, China

Jun Peng
College of Electrical and Information Engineering
Chongqing University of Science and Technology
Chongqing, China

Qiang Shen
Institute of Mathematics, Physics and Computer Science
Aberystwyth University
Aberystwyth, UK

ISBN 978-981-13-4661-3 ISBN 978-981-13-1654-8 (eBook)
https://doi.org/10.1007/978-981-13-1654-8

This Springer imprint is published by the registered company Springer Nature Singapore Pte Ltd.
The registered company address is: 152 Beach Road, #21-01/04 Gateway East, Singapore 189721, Singapore

Acknowledgements

I am very grateful for the various supports I have received during the years of carrying this research. Special thanks and gratitude particularly go to my principal supervisor, Prof. Qiang Shen, for his encouragement, enthusiasm and guidance throughout my Ph.D. process. His comments and criticisms were often pivotal in shaping the way that I considered the subject matter at hand.

I would like to thank the Department of Computer Science, Aberystwyth University. I am particularly grateful to all of the academic, administrative, technical and support staff at the department, for their kind assistance throughout my entire study and research.

Many thanks go to the current and previous members of Advanced Reasoning Group for their help, friendship, fun and support. I am especially grateful to Dr. Ren Diao for these collaborative efforts. I would also like to extend my sincere thanks and appreciation to Dr. Neil Mac Parthaláin for his advice, enthusiasm and friendship.

My sincere gratitude goes to Chongqing University of Science and Technology and Prof. Jun Peng, for their unfailing support over the years.

I would also like to thank all my friends, especially the Chinese student community in Aberystwyth for their continuous support. The work described in this book was supported by the Research Foundation of National Science and Technology Major Project (Grant No. 2016ZX05060-027), the Chongqing Municipal Key Laboratory of Institutions of Higher Education for Industrial Processes Online Analysis and Control, the Foundation of Doctor and Professor's Research Project of Chongqing University of Science and Technology (Grant No. CK2016B02), the Science and Technology Research Program of Chongqing Municipal Education Commission (Grant No. KJ1713329) and the Education Teaching Reform Research Key Project of Chongqing University of Science and Technology (Grant No. 201614).

Lastly, but by no means least, my greatest thanks go to my dear wife Fengxian Zhu and my lovely little girl Xiaoding Jin and my sister Shanghua Jin for their unreserved support and encouragement over the years. Their love has been an important part of the catalyst and driving force behind my research, and I therefore dedicate this book to them.

Contents

Acronyms

AAFMF	Applicability to arbitrary fuzzy membership functions
AC	Approximation capability
AI	Artificial intelligence
AR	Approximate reasoning
BFRI	Backward fuzzy rule interpolation
BFRIE	Backward fuzzy rule interpolation and extrapolation
CI	Computational intelligence
CNF	Convex and normal fuzzy set
CRB	Compatibility with rule base
CRI	Compositional rule of inference
EP	Extension Principle
FAR	Fuzziness of approximated result
F-BFRI	Feedback approach to backward fuzzy rule interpolation with multiple missing antecedent values
FIVE	Vague environment-based interpolation method
FIS	Fuzzy inference system
FRI	Fuzzy rule interpolation
FS	Feature selection
HBFRI	Hierarchical bidirectional fuzzy rule interpolation
HFRI	Hierarchical fuzzy rule interpolation
HFS	Hierarchical fuzzy system
IMUL	Fuzzy interpolation technique for multidimensional input spaces
IRCT	α-cut and transformation-based interpolation method
MACI	Modified α-cut-based fuzzy interpolation method
MMARS	Multiple multiantecedent rules for support
MS	Mapping similarity
P-BFRI	Parametric approach to backward fuzzy rule interpolation with multiple missing antecedent values
PIB	Preserving "In Between"
PPWL	Preservation of piecewise linearity

RP	Resolution Principle
S-BFRI	Backward fuzzy rule interpolation with single missing antecedent value
SBR	Similarity-based reasoning
ST	Similarity transfer interpolation method
T-FRI	Scare and move transformation-based fuzzy rule interpolation
TRA	Terrorism risk assessment

List of Figures

List of Tables

Abstract

Fuzzy inference, as one of the foundational principal means of approximate reasoning, has been widely used to represent and manage imprecision and vagueness in common-sense reasoning with high performance and interpretability. Fuzzy interpolation, a particular type of fuzzy inference, strengthens the power of fuzzy inference by enhancing the robustness of fuzzy systems in ensuring that a certain conclusion can always be reached if desired, when given antecedent. However, in real-world applications with interconnected rule bases, situations may arise where certain crucial antecedents are absent from given observations. If such missing antecedents are involved in the subsequent interpolation process, the final conclusion would not be deducible using conventional means.

This book presents, as its main part, a novel approach named backward fuzzy rule interpolation and extrapolation. This method allows the observations which directly relate to the conclusion to be inferred or interpolated from other antecedents and conclusion. Based on the scale and move transformation interpolation, this approach supports both interpolation and extrapolation which involve multiple hierarchical intertwined fuzzy rules, each with multiple antecedents. As such, it offers a means to broaden the application of fuzzy rule interpolation and fuzzy inference.

The work presented deals with the general situation where there may be more than one antecedent value missing for a given problem. Two techniques termed the *parametric approach* and *feedback approach* are proposed in an attempt to perform backward interpolation with multiple missing antecedent values. In addition, to further enhance the versatility and potential of backward fuzzy rule interpolation, this method is also extended to support α-cut-based interpolation by employing a fuzzy interpolation mechanism for multidimensional input spaces. Furthermore, a novel approach based on hierarchical bidirectional fuzzy interpolation is proposed in an effort to remove any inconsistencies in a rule base. Finally, from a viewpoint of integrated application analysis, experimental studies based on a real-world scenario—terrorism risk assessment—are provided in order to demonstrate the potential and efficacy of the hierarchical fuzzy rule interpolation methodology.

Chapter 1
Introduction

Approximate reasoning (AR) [1, 2] is a group of methodologies and techniques, which concentrate on the processing of inexact information containing imprecision and uncertainty in artificial intelligence (AI) and computational intelligence (CI). AR has always been committed to encoding human perceptive knowledge into numerical data such that the data can be processed by mathematical modelling, especially for systems which are too complex to be handled by human beings. The principal techniques in AR are fuzzy sets [3, 4], probabilistic reasoning [5–8], artificial neural networks [9, 10]. Each of these constituents of AR places its own emphasis on dealing with imprecision or uncertainty. In particular, fuzzy set theory offers one distinct approach to representing and dealing with uncertain knowledge and data, which lays out a main foundation of AR [3, 11]. Compared with the other principal techniques in AR, a distinguishing advantage of fuzzy sets is that they can preserve interpretability and transparency during the reasoning process due to the use of linguistic terms in fuzzy logic [12, 13].

Fuzzy set theory allows for the inclusion of vague human assessments in computing problems. Also, it provides an effective means for conflict resolution of multiple criteria and better assessment of options. Today, fuzzy sets have become an increasingly popular methodology for the modelling of various kinds of common sense reasoning, especially when dealing with nonlinear, uncertain, vague, partially true and complex systems, such as information processing [14–16], mechanical control [17–19], classification tasks [20, 21], natural language processing [22, 23], expert systems [24, 25], image recognition [26, 27], diagnosis [28–30] and intelligent decision support systems [31–33].

S. Jin et al., *Backward Fuzzy Rule Interpolation*,
https://doi.org/10.1007/978-981-13-1654-8_1

1.1 Fuzzy Logic

1.1.1 Fuzzy Sets

The theory of fuzzy logic is based on the notion of relative graded membership, as inspired by the processes of human perception and cognition [3]. This theory extends traditional set theory by softening the crisp or discrete boundaries of classical sets. In classical set theory, elements can merely either fully belong to a certain set or not belong to it at all. For example, when describing the concept *high temperature*, defined here over the domain [25 °C, 50 °C], the element 37 °C is classified as high temperature, whereas the element 20 °C is not. However, once the elements are included in the same set, no distinction is made between them; all elements are treated as belonging to the set equally. By definition, let $\mathbb{U}$ be the universe of discourse and x an element of $\mathbb{U}$. A classical set $A \subseteq \mathbb{U}$ is defined as a collection of ordered pairs (e.g. $(x, 0)$ or $(x, 1)$, where $x \in \mathbb{U}$) for each element. More specifically, $(x, 0)$ indicates that $x \notin A$, whereas $(x, 1)$ stands for $x \in A$.

In the above example, 37 °C is higher than 30 °C, even if they may be both considered to be *high temperature* in classical set theory. Thus, under many circumstances, such clear cut distinctions in classical sets may result in the loss of useful information. In order to combat this, crisp sets are further extended by fuzzy set theory, which allows elements to belong to multiple sets with partial degrees. In contrast to classical sets, fuzzy sets relax the crisp boundary by allowing memberships to take any value within the range of [0, 1]. Given a universe of discourse $\mathbb{U}$, a fuzzy set A is defined as a set of ordered pairs:

$$A = \{(x, \mu_A(x)) | x \in \mathbb{U}, \mu_A(x) \in [0, 1]\} \tag{1.1}$$

where $\mu_A(x)$ is termed the membership function of A, mapping each element in the universe of discourse $\mathbb{U}$ to a membership grade ranging from 0 to 1.

Returning to the example, it may be better to represent the concept *high temperature* as a fuzzy set. An example membership function for this set can be found in Fig. 1.1, defined over a range of temperature values. Given that the value of the measurement is 37 °C, it can be determined that this temperature belongs to the set *high temperature* with a membership degree of 0.80. Similarly, if the temperature is 30 °C, then the resulting degree of membership is 0.33. Here, both 30 and 37 °C partially belong to the fuzzy set *high temperature*, but 37 °C has a higher degree of membership to this set. Thus, fuzzy sets provide an efficient means of distinguishing elements within the same set. In the literature, many different membership functions have been defined with various shapes. The most common are: *triangular, trapezoidal, singleton and Gaussian* membership functions. Typical examples of these are illustrated in Fig. 1.2.

Two frequently used membership functions for real-world application domains, trapezoidal and triangular [34, 35], are also employed in this work and are defined in the following forms:

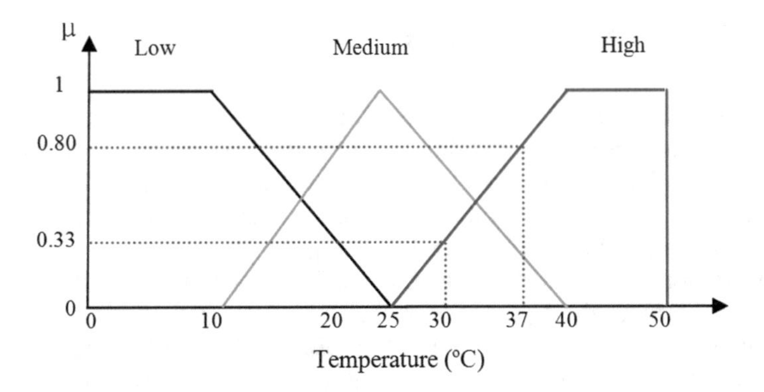

Fig. 1.1 Definition of fuzzy sets for temperature

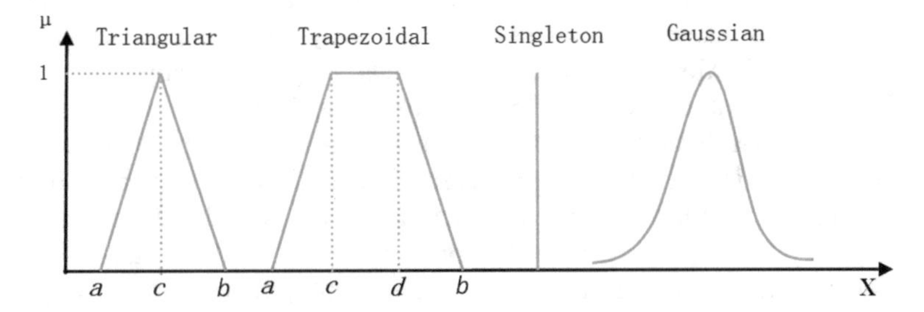

Fig. 1.2 Examples of fuzzy membership functions

- Triangular membership:

$$\mu(x) = \begin{cases} 0, & if \;\; x < a \\ \dfrac{x-a}{c-a}, & if \;\; x \in [a, c] \\ \dfrac{b-x}{b-c}, & if \;\; x \in [c, b] \\ 0, & if \;\; x > b \end{cases} \tag{1.2}$$

  where c is the central element which has a full membership value; a and b denote the lower and upper bounds, respectively.
- Trapezoidal membership:

$$\mu(x) = \begin{cases} 0, & if \;\; x < a \\ \dfrac{x-a}{c-a}, & if \;\; x \in [a, c] \\ 1, & if \;\; x \in [c, d] \\ \dfrac{b-x}{b-d}, & if \;\; x \in [d, b] \\ 0, & if \;\; x > b \end{cases} \tag{1.3}$$

1.1.2 Fuzzy Inference Systems

Fuzzy inference systems (FIS) have been increasingly applied to real-world problems. This is because of their ability to mimic the intuitive ability of the human mind to summarise data and focus on decision-relevant information. Traditional FIS consists of four interconnected processes [19]. The basic overview of a typical fuzzy inference system is illustrated in Fig. 1.3: firstly, the fuzzifier component takes the crisp input numbers and converts them into a fuzzy input using input membership functions to calculate the degree of truth related to each input. Afterwards, the inference engine part takes the fuzzy input from the fuzzifier and applies it to a rule base. Then, the membership functions are output to create a fuzzy output. Lastly, the defuzzifier converts the output fuzzy sets into certain required numerical outputs in order to make informed decisions. For certain applications, such as [36, 37], the use of linguistic terms that are themselves defined as fuzzy sets is employed directly. In these cases, fuzzification and defuzzification are no longer necessary.

Rules, which are at the heart of FIS for representing domain knowledge, are coupled with membership functions in which each rule can be thought of as a subsystem. Rules themselves do nothing unless inputs are applied to them. Fuzzy rules or fuzzy conditional statements are expressions of the form IF A THEN B, where A and B are labels of fuzzy sets [3] characterised by appropriate membership functions. Due to their concise form, fuzzy IF-THEN rules are often employed to capture the imprecise forms of reasoning that play an essential role in the human ability to make decisions in an environment of vagueness.

There are two main ways in which to construct a rule base for a given problem. The first is by directly translating expert knowledge into rules, and fuzzy inference systems with such rule bases are usually called fuzzy expert systems [37]. Because rules are fuzzy representations of *expert* knowledge, rule bases offer a high level of semantic interpretability and good generalisation capability. However, for complex systems, it is difficult to build a rule base in this way, which has led to an alternative approach for rule base construction. This approach is data-driven, and fuzzy rules are obtained from data using machine learning techniques rather than direct expert

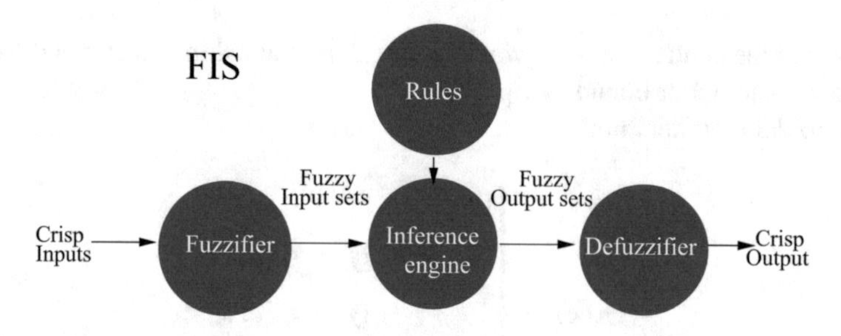

Fig. 1.3 Typical conventional single layer fuzzy inference system

knowledge [38, 39]. In contrast, rule bases built in this way lack comprehensibility and transparency.

Given a fuzzy rule base generated in either of the above two ways, there are a number of fuzzy inference mechanisms that can be used to derive a consequent from a given observation. The two most important models are the compositional rule of inference (CRI) and similarity-based reasoning (SBR), which are introduced in the following two subsections, respectively.

1.1.2.1 Compositional Rule of Inference

CRI is a fuzzy extension of classical *modus ponens* which can be viewed from two different perspectives. Firstly, classical *modus ponens* only supports predicates concerning singleton elements, but CRI is able to deal with predicates which concern a set of elements in variable domains. This is achieved by representing a fuzzy rule as a fuzzy relation over the Cartesian product of the domains of the antecedent and consequent variables. Various fuzzy implication relations have been proposed [40–43], each of which may have its own properties and therefore render them suitable only for a certain group of applications or problems. Secondly, classical *modus ponens* only supports crisp/discrete logic, whereas CRI supports multi- or continuously valued logic. That is, CRI is able to deal with predicates with partial truth values, which are implemented by a compositional operator supremum T, where T represents a t-norm [44]. t-norm is a binary function $T : [0, 1] \times [0, 1] \rightarrow [0, 1]$ on the interval $[0, 1]$ satisfying the following conditions:

$$\begin{aligned} T(x, y) &= T(y, x) \quad (commutativity) \\ T(x, T(y, z)) &= T(T(x, y), z) \quad (associativity) \\ T(x, y) \leq T(x, z) \quad whenever \quad y &\leq z \quad (monotonicity) \\ T(x, 1) &= x \quad (neutral \;\; element \;\; 1) \end{aligned} \tag{1.4}$$

There are two very important features of the compositional rule of inference under t-norms: (1) if the t-norm defining the composition and the membership function of the given observation is both continuous, then the conclusion depends continuously on the observation; (2) if there is a fuzzy relation over the Cartesian product of the domains of the antecedent and consequent variables, and the t-norm and the membership function of the relation are continuous, then the observation has a continuous membership function.

A number of existing fuzzy reasoning methods based on CRI have been proposed in the literature [45, 46]. In particular, the first successful practical approach was Mamdani inference [47] which is also commonly implemented fuzzy methodology in physical control systems at present [48, 49]. It was originally proposed as an attempt to control a steam engine and boiler combination by synthesising a set of linguistic control rules obtained from experienced human operators. Mamdani inference implements CRI by using *minimum* as the t-norm operator due to its simplicity.

In particular, the inferred result from each fired rule is a fuzzy set which is transformed from the rule consequence by restricting the membership of those elements whose memberships are greater than the firing strength. The firing strength is also sometimes termed the *degree of confidence or satisfaction*, which is the supremum within the variable domain of the minimum of the rule antecedent and the given observation. Mamdani controllers can be further simplified by removing the defuzzification phase [50]. This enhances the efficiency of fuzzy controllers because it greatly simplifies the computation required by the more general Mamdani method.

Another approach is called Takagi–Sugeno–Kang (TSK) inference. The basic idea of TSK method is to decompose the input space into fuzzy regions and to approximate the system in every region by a simple mathematical model. The overall fuzzy model is thus considered to be a combination of interconnected subsystems with simpler models. TSK has fuzzy sets involved only in the premise part, which can be applied to the fuzzy dynamic models. The great advantage of the TSK model is its representative power; it is capable of describing a highly nonlinear system using a small number of rules. Moreover, the output of the TSK model has an explicit functional expression form; it is convenient to identify its parameters using some learning algorithms. Although these algorithms can generate a TSK model with good global performance (i.e. the model is capable of approximating a given system with arbitrary accuracy, provided that sufficient rules are used and sufficient training data are available), they cannot guarantee that the resulting model will have good local performance. Often, the submodels in the TSK model may exhibit erratic local behaviour, which is difficult to interpret [51, 52]. So far, several relevant methods [38, 52–54] have been proposed to address the interpretability issue of the TSK model.

1.1.2.2 Similarity-Based Fuzzy Inference

Despite the success of CRI in various fuzzy system applications, it suffers from various shortcomings including its complexity and its unclear underlying semantic interpretability of the logical implication [42, 55]. This has led to another group of fuzzy reasoning approaches which are based on the concept of similarity. This approach is usually called similarity-based fuzzy reasoning [56–61]. Similarity considerations play a major role in human cognitive processes [62] and, hence, also in approximate reasoning. It is intuitive that if a given observation is similar to the antecedent of a rule, the conclusion from the observation should also be similar to the consequence of the rule. In contrast to CRI-based fuzzy reasoning, similarity-based fuzzy reasoning does not require the construction of a fuzzy relation. Instead, it is based on the degree of similarity (with respect to a given similarity metric) between the given observation and the antecedent of a rule. By using the computed similarity degree, the consequence of the fired rule can be modified to the consequence of the given observation.

A similarity-based fuzzy inference approach, approximate analogical reasoning schema, was proposed in [42, 55]. In this method, rules are fired according to the similarity degrees between a given observation and the antecedents of rules. If the degree of similarity between the given observation and the antecedent of a rule is greater than a predefined threshold value, the rule will be fired and the consequence of the observation is deduced from the rule consequence by a prespecified modification procedure. Another similarity-based fuzzy inference approach was proposed in [58, 63], which particularly targets medical diagnostic applications. This approach is based on the *cosine* angle between the two vectors that represent the actual and the user's specified values of the antecedent variable. Several other methods that use different modification procedures can be found in [64]. In particular, a fuzzy reasoning technique which employs similarity measures based on the degree of subsethood between the propositions in the antecedent and a given observation is proposed in [65]. This method has also been extended to consider the weights of the propositions in the antecedent [61, 66]. Similarity-based fuzzy inference approaches usually arrive at solutions with more natural appeal than those introduced in the last subsection, but further research is required in order to determine which particular approach in this group performs best.

1.2 The "Curse of Dimensionality"

As the application domain of fuzzy systems expands to more complex ones, a serious limitation of the conventional fuzzy systems became apparent: it was discovered that the number of rules in a standard fuzzy system increases exponentially with the number of variables involved. Suppose that a fuzzy model contains K variables and each variable is partitioned into M fuzzy values. The order of the number of rules in the rule base is therefore $O(M^K)$. This is usually referred to as the rule-explosion problem or curse of dimensionality.

To overcome the "curse of dimensionality", several methods have been proposed for optimising the size of the rule base obtained using automated modelling techniques [67–69]. Similarity-driven rule base simplification [70] differs from other reduction methods (such as fuzzy clustering [13, 71]) in that its main objective is to reduce the number of fuzzy sets used in the model. It does not necessarily alter the number of rules. Reduction of the number of rules may follow from rule base simplification if the rules become equal as a result of the merging process. If no rules are combined, simplification can still be achieved by reducing the number of fuzzy sets.

Another way to deal with this problem is through the use of hierarchical fuzzy systems (HFS) [16, 72–74]. HFS can improve transparency and interpretability in many high-dimensional situations. A hierarchical fuzzy system consists of a number of hierarchically connected low-dimensional fuzzy systems. The number of rules in the hierarchical fuzzy system increases linearly with an increasing number of input variables. In conventional (non-hierarchical) fuzzy systems, suppose that there are K input variables and M membership functions for each variable, then M^K rules are needed in order to construct a system that fully covers the underlying domain. An K input hierarchical fuzzy system comprises $K - 1$ low-dimensional fuzzy systems, if

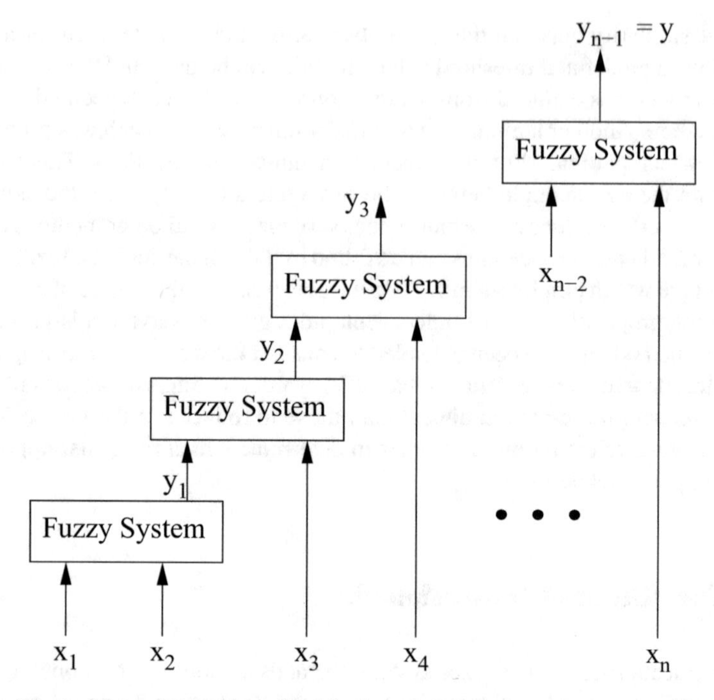

Fig. 1.4 A typical example of a hierarchical fuzzy system

each subsystem has two inputs. In this case, given M fuzzy sets for each variable, the total number of rules is $(K - 1)M^2$ which is a linear function of the number of input variables. A typical example of hierarchical fuzzy systems is shown in Fig. 1.4 [74].

1.3 Fuzzy Rule Interpolation (FRI)

Reducing the number of fuzzy terms K for each variable may result in a sparse fuzzy rule base. More importantly, in real-world applications, it is difficult to obtain sufficient data with input variables that cover the whole input space. There are also other situations which may result in the generation of sparse fuzzy rule bases:

(a) The performance of fuzzy expert systems is often criticised when the number of inputs is large. The size of the rule base and the complexity grow exponentially with the number of inputs, which leads to a lack of interpretability. One possible solution to reducing the complexity is the elimination of redundant rules. However, this often leads to sparse rule bases [75].
(b) Incomplete knowledge about the system which is being modelled can result in sparse rule bases [76]. In fuzzy expert systems constructed using linguistic rules,

an incomplete rule base can be a consequence of missing expertise for a certain system configuration or state space regions. On the other hand, if the systems are generated from numerical sample data, a situation may occur where sample data do not sufficiently represent some regions of the input domain.

(c) Tuning the original dense rule bases by learning usually leads to typical, frequently occurring situations, which are represented by the majority of the training data/expert knowledge. Therefore, rule antecedents can be partially shifted and "shrunk", which means that the tuned model may also contain gaps and, hence, sparse rule bases [77].

(d) "Gaps" can be defined between rule bases intentionally, in order to avoid high complexity in very large systems. Hence, fuzzy interpolation techniques have an important role to play in hierarchically structured systems [78].

When a given observation has no overlap with antecedent values, no rule can be invoked in classical fuzzy inference, and therefore, no consequence can be derived. This problem is usually referred to as the tomato problem [79]. To address this problem, fuzzy rule interpolation (FRI) was originally proposed in [80, 81]. It is of particular significance for reasoning in the presence of insufficient knowledge or sparse rule bases. The techniques of FRI may not only support inference in such situations, but also help to reduce the complexity of fuzzy models. A number of important interpolation approaches have been proposed in the literature, including [75, 82–93].

1.4 Backward Fuzzy Rule Interpolation (BFRI)

Despite the proposal of numerous FRI approaches, there are relatively few examples of practical applications [94]. One of the main reasons for this is that many applications involve multiple-input and multiple-output problems. The rules are typically irregular in nature (i.e. not always using the same antecedents). In particular, rules may be arranged in an interconnected mesh, where observations and conclusions in between different subsets of rules may be overlapping, and yet not directly related throughout the entire rule base. For such complex systems, any missing values in a given set of observations may lead to failure in interpolation. Therefore, the interpolation of such crucial missing values may become necessary, in order to facilitate inference.

To address such problems, the work in this book proposes a novel approach termed backward fuzzy rule interpolation (BFRI). This approach enables unknown antecedent values to be interpolated, given other antecedents and the conclusions. The proposed techniques support flexible interpolation when certain antecedents are missing from the observation, where traditional FRI methods fail. In addition, BFRI also enables indirect interpolative reasoning, which involves several fuzzy rules, each with multiple antecedents. Therefore, it offers a means to broaden the application of fuzzy rule interpolation and fuzzy inference.

For many application problems, however, there are often more than one missing antecedent value, rather than just one unknown antecedent. This, therefore, poses the question: how can BFRI be performed with multiple missing values? To address this challenging issue, two methods are developed. The first directly extends the single missing antecedent case, by computing and searching for good-quality parameter combinations for the scale and move transformation-based interpolation process. The second approach works more closely with conventional FRI procedures, by estimating the possible missing antecedent values and subsequently verifying the interpolative outcome against the observation.

To enhance the versatility and flexibility of the BFRI concept, the backward fuzzy interpolation method is also extended to support the α-cut-based interpolation method in this book. (The α-cut A_α ($\alpha \in (0, 1]$) of a fuzzy set A is a crisp set; it contains the elements of the universe of discourse with membership degree not smaller than α; formally, it is defined as: $A_\alpha = \{x|A(x) \geq \alpha, \alpha \in (0, 1]\}$.) The fuzzy interpolation technique for multidimensional input spaces (IMUL) method is adopted in order to achieve this. Furthermore, a hierarchical bidirectional fuzzy interpolation (HBFRI) methodology is proposed, which can provide a flexible solution to hierarchical fuzzy inference system. Finally, HBFRI is implemented for a real-world application: terrorism risk assessment in order to demonstrate its potential and efficacy.

1.5 Book Structure

The remainder of this book is structured as follows:

- Chapter 2: **Background: Fuzzy Rule Interpolation**
  This chapter presents an overview of existing fuzzy interpolation approaches and lays out the foundation and motivation for the current work. FRI is categorised into two principal groups: single-step fuzzy interpolation and intermediate rule-based fuzzy interpolation. Each group is associated with a detailed description of three typical methods as well as their extensions and properties. In particular, the scale and move transformation-based interpolation procedure is introduced in detail, and its underlying key concepts and essential formula are defined, which will be used as the main technology to implement backward fuzzy rule interpolation methodology in this book. A brief summary of the evaluation of fuzzy interpolation approaches ends the chapter.
- Chapter 3: **Transformation-Based BFRI with Single Missing Antecedent Value**
  This chapter details one of the key contributions of the book: the BFRI concept and associated interpolation algorithm for a single missing antecedent value involving multiple multi-antecedent rules. This is a novel FRI approach based on reverse reasoning. For completeness, the general concept of BFRI and its key notions are first provided. A method, named S-BFRI, is proposed for interpolation involving situations where the consequent value is known and the values of all but one antecedent variable are also given. The task is to estimate the value of that single unknown

antecedent. This chapter also presents worked examples that demonstrate the performance of S-BFRI. The results of BFRI are verified by performing conventional FRI, using the reconstructed observation involving the obtained missing value. The work in this chapter has been published in [95, 96], which have attracted broad interest (with [96] receiving the best paper award at the 2012 IEEE International Conference on Fuzzy Systems).

- Chapter 4: **Transformation-Based BFRI with Multiple Missing Antecedent Values**
  This chapter presents another key contribution of the book: BFRI with multiple missing antecedent values, which may be more common in practical problems such as medical diagnosis, network intrusion detection and oil exploration. The use of this technique, termed M-BFRI, allows better reasoning performance than that achievable with the single missing antecedent method. Two approaches are presented in order to deal with the situations where multiple antecedent values are missing from a given observation: the P-BFRI approach and the F-BFRI approach. The former directly extends the S-BFRI method but involves a higher computational complexity; the later is a more generalised method that works more closely with conventional FRI procedures. Worked examples are given to demonstrate the performance of both P-BFRI and F-BFRI. Systematic experimental comparative studies are given also. Both approaches have an acceptable accuracy and computational complexity when compared with the state of the art. The techniques described in this chapter have been published in [97, 98].
- Chapter 5: **An Alternative Backward Fuzzy Rule Interpolation Method**
  In order to strengthen the versatility and feasibility of backward fuzzy interpolative reasoning, in this chapter, an alternative α-cut-based interpolation method is proposed. In particular, the fuzzy interpolation technique for multidimensional input spaces (IMUL) allows interpolation using the rules with multidimensional input spaces. It also guarantees that the interpolative outcomes are crisp if the inputs (observations) are crisp. IMUL is employed to extend the existing BFRI. Numerical examples and comparative studies are provided to demonstrate the efficacy of this work. The techniques described in this chapter have been published in [99].
- Chapter 7: **Application: Terrorism Risk Assessment using BFRI**
  Terrorist attacks launched by extremist groups or individuals have caused catastrophic consequences worldwide. Terrorism risk assessment therefore plays a crucial role in national and international security. The proposed BFRI approach is employed in this chapter to build a hierarchical bidirectional fuzzy rule interpolation methodology. HBFRI is particularly effective in situations where a multiple multi-antecedent rule-based system needs to be reconstructed into a multilayered fuzzy system, and where each of the layers are themselves individual sparse rule-based systems.
  HBFRI offers not only a significant reduction in the number of rules, but also flexibility of interpolation for situations where certain crucial antecedents are absent (or assumed to be missing) from given observations. Structured assessments and realistic case studies confirm the effectiveness of the developed work when applied to a given terrorist attack prediction or decision support scenarios.

The proposed methodology shows its strength in developing a flexible and comprehensive response to terrorism threats. The work described in this chapter is currently under review for potential journal publication [100].

- Chapter 8: **Conclusion**
  This chapter summarises the key contributions made by the book, together with a discussion of topics which form the basis for future research.

References

1. Q. Shen, R. Leitch, Fuzzy qualitative simulation. IEEE Trans. Syst. Man Cybern. **23**(4), 1038–1061 (1993)
2. Q. Shen, R. Leitch, Fuzzy logic and approximate reasoning. Synthese **30**(3–4), 407–428 (1975)
3. L. Zadeh, Fuzzy sets. Inf. Control **8**(3), 338–353 (1965)
4. L. Zadeh, Quantitative fuzzy semantics. Inf. Sci. **3**(2), 159–176 (1971)
5. G.F. Cooper, The computational complexity of probabilistic inference using bayesian belief networks. Artif. Intell. **42**(2), 393–405 (1990)
6. D. Heckerman, D. Geiger, D.M. Chickering, Learning bayesian networks: the combination of knowledge and statistical data. Mach. Learn. **20**(3), 197–243 (1995)
7. R.E. Neapolitan, *Probabilistic Reasoning in Expert Systems: Theory and Algorithms* (CreateSpace Independent Publishing Platform, 2012)
8. J. Pearl, *Probabilistic Reasoning in Intelligent Systems: Networks of Plausible Inference* (Morgan Kaufmann, 1988)
9. C.M. Bishop, *Neural Networks for Pattern Recognition* (Oxford University Press, 1995)
10. S. Haykin, N. Network, A comprehensive foundation. Neural Netw. **2**, 2004 (2004)
11. D. Neagu, V. Palade, A neuro-fuzzy approach for functional genomics data interpretation and analysis. Neural Comput. Appl. **12**(3–4), 153–159 (2003)
12. T.J. Ross, *Fuzzy Logic with Engineering Applications* (Wiley, 2009)
13. M. Sugeno, T. Yashukawa, A fuzzy-logic-based approach to qualitative modeling. IEEE Trans. Fuzzy Syst. **1**(1), 7–31 (1993)
14. T. Martin, Fuzzy sets in the fight against digital obesity. Fuzzy Sets Syst. **156**(3), 411–417 (2005)
15. L.-X. Wang, J.M. Mendel, Fuzzy basis functions, universal approximationand orthogonal least-squares learning. IEEE Trans. Neural Netw. **3**(5), 807–814 (1992)
16. X.-J. Zeng, J.A. Keane, Approximation capabilities of hierarchical fuzzy systems. IEEE Trans. Fuzzy Syst. **13**, 659–672 (2005)
17. J.J. Buckley, Sugeno type controllers are universal controllers. Fuzzy Sets Syst. **53**(3), 299–303 (1993)
18. J.L. Castro, Fuzzy logic controllers are universal approximators. IEEE Trans. Syst. Man Cybern. **25**(4), 629–635 (1995)
19. C.-C. Lee, Fuzzy logic in control systems: fuzzy logic controller I. IEEE Trans. Syst. Man Cybern. **20**(2), 404–418 (1990)
20. P.P. Angelov, X. Zhou, Evolving fuzzy-rule-based classifiers from data streams. IEEE Trans. Fuzzy Syst. **16**(6), 1462–1475 (2008)
21. D. Nauck, R. Kruse, A neuro-fuzzy method to learn fuzzy classification rules from data. Fuzzy Sets Syst. **89**(3), 277–288 (1997)
22. H.M. Hersh, A. Caramazza, A fuzzy set approach to modifiers and vagueness in natural language. J. Exp. Psychol. Gen. **105**(3), 254 (1976)
23. F. Wang, Towards a natural language user interface: an approach of fuzzy query. Int. J. Geogr. Inf. Sci. **8**(2), 143–162 (1994)

24. F. Forsyth, *Expert Systems Principles* (Chapman & Hall Ltd., 1984)
25. M. Schneider, A. Kandel, G. Langholz, G. Chew, *Fuzzy Expert System Tools* (Wiley, 1996)
26. W. Cai, S. Chen, D. Zhang, Fast and robust fuzzy c-means clustering algorithms incorporating local information for image segmentation. Pattern Recogn. **40**(3), 825–838 (2007)
27. Z. Chi, H. Yan, T. Pham, *Fuzzy Algorithms: With Applications to Image Processing and Pattern Recognition*, vol. 10 (World Scientific, 1996)
28. L. Kuncheva, F. Steimann, Fuzzy diagnosis. Artif. Intell. Med. **16**(2), 121–128 (1999)
29. H.-T. Yang, C.-C. Liao, Adaptive fuzzy diagnosis system for dissolved gas analysis of power transformers. IEEE Trans. Power Deliv. **14**(4), 1342–1350 (1999)
30. J.F.-F. Yao, J.-S. Yao, Fuzzy decision making for medical diagnosis based on fuzzy number and compositional rule of inference. Fuzzy Sets Syst. **120**(2), 351–366 (2001)
31. E.W. Ngai, F. Wat, Fuzzy decision support system for risk analysis in e-commerce development. Decis. Support Syst. **40**(2), 235–255 (2005)
32. H. Nokhbatolfoghahaayee, M.B. Menhaj, M. Shafiee, Fuzzy decision support system for crisis management with a new structure for decision making. Expert Syst. Appl. **37**(5), 3545–3552 (2010)
33. D. Petrovic, Y. Xie, K. Burnham, Fuzzy decision support system for demand forecasting with a learning mechanism. Fuzzy Sets Syst. **157**(12), 1713–1725 (2006)
34. A. Barua, L.S. Mudunuri, O. Kosheleva, Why trapezoidal and triangular membership functions work so well: towards a theoretical explanation. (Departmental Technical Reports (CS) of University of Texas at El Paso, 2013). Paper 783
35. R.E. Giachetti, R.E. Young, Analysis of the error in the standard approximation used for multiplication of triangular and trapezoidal fuzzy numbers and the development of a new approximation. Fuzzy Sets Syst. **91**(1), 1–13 (1997)
36. G. Feng, A survey on analysis and design of model-based fuzzy control systems. IEEE Trans. Fuzzy Syst. **14**(5), 676–697 (2006)
37. E.H. Mamdani, S. Assilian, An experiment in linguistic synthesis with a fuzzy logic controller. Int. J. Man Mach. Stud. 7 (1975)
38. S.E. Papadakis, J. Theocharis, A GA-based fuzzy modeling approach for generating TSK models. Fuzzy Sets Syst. **131**(2), 121–152 (2002)
39. S.E. Papadakis, J. Theocharis, Generating fuzzy rules by learning from examples. IEEE Trans. Syst. Man Cybern. B **22**(6), 1414–1427 (1992)
40. W. Bandler, L. Kohout, Fuzzy power sets and fuzzy implication operators. Fuzzy Sets Syst. **4**(1), 13–30 (1980)
41. I.B. Turksen, Four methods of approximate reasoning with interval-valued fuzzy sets. Int. J. Approx. Reason. **3**(2), 121–142 (1989)
42. I. Turksen, Z. Zhong, An approximate analogical reasoning approach based on similarity measures. IEEE Trans. Syst. Man Cybern. **18**(6), 1049–1056 (1988)
43. R.R. Yager, An approach to inference in approximate reasoning. Int. J. Man Mach. Stud. **13**(3), 323–338 (1980)
44. E.P. Klement, R. Mesiar, E. Pap, Triangular norms: position paper I: basic analytical and algebraic properties. Fuzzy Sets Syst. **143**(1), 5–26 (2004)
45. M. Mizumoto, H.-J. Zimmermann, Comparison of fuzzy reasoning methods. Fuzzy Sets Syst. **8**(3), 253–283 (1982)
46. H. Nakanishi, I. Turksen, M. Sugeno, A review and comparison of six reasoning methods. Fuzzy Sets Syst. **57**(3), 257–294 (1993)
47. E.H. Mamdani, Application of fuzzy algorithms for control of a simple dynamic plant. Proc. Inst. Electr. Eng. **121**, 1585–1588 (1974)
48. J. Canada-Bago, J. Fernandez-Prieto, M. Gadeo-Martos, J. Velasco, A new collaborative knowledge-based approach for wireless sensor networks. Sensors (Basel, Switzerland) **10**(6), 6044–6062 (2010)
49. C. Santos, F. Espinosa, D. Pizarro, F. ValdeÌĄs, E. Santiso, I. DiÌĄaz, Fuzzy decentralized control for guidance of a convoy of robots in non-linear trajectories, in *2010 IEEE Conference on Emerging Technologies and Factory Automation (ETFA)* (2010), pp. 1–8

50. B. Moser, M. Navara, Fuzzy controllers with conditionally firing rules. IEEE Trans. Fuzzy Syst. **10**(3), 340–348 (2002)
51. J. Casillas, *Interpretability Issues in Fuzzy Modeling*, vol. 128 (Springer, 2003)
52. J. Yen, L. Wang, C.W. Gillespie, Improving the interpretability of TSK fuzzy models by combining global learning and local learning. IEEE Trans. Fuzzy Syst. **6**(4), 530–537 (1998)
53. L.J. Herrera, H. Pomares, I. Rojas, O. Valenzuela, A. Prieto, Tase, a taylor series-based fuzzy system model that combines interpretability and accuracy. Fuzzy Sets Syst. **153**(3), 403–427 (2005)
54. F. Hoffmann, D. Schauten, S. Holemann, Incremental evolutionary design of TSK fuzzy controllers. IEEE Trans. Fuzzy Syst. **15**(4), 563–577 (2007)
55. F. Hoffmann, D. Schauten, S. Holemann, An approximate analogical reasoning schema based on similarity measures and interval-valued fuzzy sets. Fuzzy Sets Syst. **34**(3), 323–346 (1990)
56. M. Azzeh, D. Neagu, P.I. Cowling, Fuzzy grey relational analysis for software effort estimation. Empir. Softw. Eng. **15**(1), 60–90 (2010)
57. M. Azzeh, D. Neagu, P.I. Cowling, Analogy-based software effort estimation using fuzzy numbers. J. Syst. Softw. **84**(2), 270–284 (2011)
58. S.-M. Chen, A new approach to handling fuzzy decision-making problems. IEEE Trans. Syst. Man Cybern. **18**(6), 1012–1016 (1988)
59. M.-G. Chun, A similarity-based bidirectional approximate reasoning method for decision-making systems. Fuzzy Sets Syst. **117**(2), 269–278 (2001)
60. S. Raha, N.R. Pal, K.S. Ray, Similarity-based approximate reasoning: methodology and application. IEEE Trans. Syst. Man Cybern. Part A Syst. Hum. **32**(4), 541–547 (2002)
61. S. Raha, N.R. Pal, K.S. Ray, Improved fuzzy knowledge representation and rule evaluation using fuzzy petri nets and degree of subsethood. Int. J. Intell. Syst. **9**(12), 1083–1100 (1994)
62. R.L. Goldstone, D.L. Medin, D. Gentner, Relational similarity and the nonindependence of features in similarity judgments. Cogn. Psychol. **23**(2), 222–262 (1991)
63. S.-M. Chen, Y.-K. Ko, Y.-C. Chang, J.-S. Pan, Weighted fuzzy interpolative reasoning based on weighted increment transformation and weighted ratio transformation techniques. IEEE Trans. Fuzzy Syst. **17**(6), 1412–1427 (2009)
64. D. Yeung, E. Tsang, Fuzzy knowledge representation and reasoning using petri nets. Expert Syst. Appl. **7**(2), 281–289 (1994)
65. B. Kosko, *Neural Networks and Fuzzy Systems: A Dynamical Systems Approach to Machine Intelligence/Book and Disk*, vol. 1 (Prentice Hall, 1992)
66. D. Yeung, E. Ysang, A multilevel weighted fuzzy reasoning algorithm for expert systems. IEEE Trans. Syst. Man Cybern. B **28**(2), 149–158 (1998)
67. U. Kaymak, R. Babuska, Compatible cluster merging for fuzzy modeling, in *Proceedings of the FUZZ-IEEE/IFES95* (1995), pp. 897–904
68. B. Song, R. Marks, S. Oh, P. Arabshahi, T. Caudell, J. Choi et al., Adaptive membership function fusion and annihilation in fuzzy if-then rules, in *Second IEEE International Conference on Fuzzy Systems, IEEE* (1993), pp. 961–967
69. C. Sun, Rule-base structure identification in an adaptive-network-based fuzzy inference system. IEEE Trans. Fuzzy Syst. **2**(1), 64–73 (1994)
70. M.L.F. Herrera, J. Verdegay, Tackling real-coded genetic algorithm: operators and tools for behavioral analysis. Artif. Intell. Rev. **12**, 265–319 (1998)
71. J.C. Bezdek, *Pattern Recognition with Fuzzy Objective Function* (Plenum, New York, 1981)
72. G. Raju, J. Zhou, Adaptive hierarchical fuzzy controller. IEEE Trans. Syst. Man Cybern. **23**(4), 973–980 (1993)
73. G. Raju, J. Zhou, A. Roger, Hierarchical fuzzy control. Int. J. Control **54**(5), 1201–1216 (1991)
74. L. Wang, Universal approximation by hierarchical fuzzy systems. Fuzzy Sets Syst. **93**(2), 223–230 (1998)
75. L. Wang, Size reduction by interpolation in fuzzy rule bases. IEEE Trans. Syst. Man Cybern. B **27**(1), 14–25 (1997)

76. K.W. Wong, D. Tikk, T.D. Gedeon, L.T. Kóczy, Fuzzy rule interpolation for multidimensional input spaces with applications: a case study. IEEE Trans. Fuzzy Syst. **13**(6), 809–819 (2005)
77. D.G. Burkhardt, P.P. Bonissone, Automated fuzzy knowledge base generation and tuning, in *IEEE International Conference on Fuzzy Systems, IEEE* (1992), pp. 179–188
78. L. Kóczy, K. Hirota, Approximate inference in hierarchical structured rule bases, in *Proceedings of 5th IFSA World Congress (IFSA93)* (1993), pp. 1262–1265
79. L. Kóczy, K. Hirota, Comparison of fuzzy reasoning methods. Fuzzy Sets Syst. **8**(3), 253–283 (1982)
80. L. Koczy, K. Hirota, Approximate reasoning by linear rule interpolation and general approximation. Int. J. Approx. Reason. **9**(3), 197–225 (1993)
81. L. Koczy, K. Hirota, Interpolative reasoning with insufficient evidence in sparse fuzzy rule bases. Inf. Sci. **71**(1–2), 169–201 (1993)
82. P. Baranyi, L.T. Kóczy, T.D. Gedeon, A generalized concept for fuzzy rule interpolation. IEEE Trans. Fuzzy Syst. **12**(6), 820–837 (2004)
83. S. Chen, Y. Ko, Fuzzy interpolative reasoning for sparse fuzzy rule-based systems based on α-cuts and transformations techniques. IEEE Trans. Fuzzy Syst. **16**(6), 1626–1648 (2008)
84. D. Dubois, H. Prade, On fuzzy interpolation*. Int. J. Gen. Syst. **28**(2–3), 103–114 (1999)
85. W. Hsiao, S. Chen, C. Lee, A new interpolative reasoning method in sparse rule-based systems. Fuzzy Sets Syst. **93**(1), 17–22 (1998)
86. Z. Huang, Q. Shen, Fuzzy interpolative reasoning via scale and move transformations. IEEE Trans. Fuzzy Syst. **14**(2), 340–359 (2006)
87. Z. Huang, Q. Shen, Fuzzy interpolation and extrapolation: a practical approach. IEEE Trans. Fuzzy Syst. **16**(1), 13–28 (2008)
88. L.T. Kóczy, K. Hirota, L. Muresan, Interpolation in hierarchical fuzzy rule bases, in *Proceedings of International Conference on Fuzzy Systems* (2000), pp. 471–477
89. D. Tikk, P. Baranyi, Comprehensive analysis of a new fuzzy rule interpolation method. IEEE Trans. Fuzzy Syst. **8**(3), 281–296 (2000)
90. Y. Yam, L. Kóczy, Representing membership functions as points in high-dimensional spaces for fuzzy interpolation and extrapolation. IEEE Trans. Fuzzy Syst. **8**(6), 761–772 (2000)
91. Y. Yam, M. Wong, P. Baranyi, Interpolation with function space representation of membership functions. IEEE Trans. Fuzzy Syst. **14**(3), 398–411 (2006)
92. Y. Yam, M. Wong, P. Baranyi, Adaptive fuzzy interpolation. IEEE Trans. Fuzzy Syst. **19**(6), 1107–1126 (2011)
93. Y. Yam, M. Wong, P. Baranyi, Closed form fuzzy interpolation. Fuzzy Sets Syst. **225**, 1–22 (2013)
94. S. Kovács, Special issue on fuzzy rule interpolation. J. Adv. Comput. Intell. Intell. Inform. 253 (2011)
95. S. Jin, R. Diao, Q. Shen, Towards backward fuzzy rule interpolation, in *Proceedings of the 11th UK Workshop on Computational Intelligence (UKCI2011)* (2011), pp. 194–200
96. S. Jin, R. Diao, Q. Shen, Backward fuzzy interpolation and extrapolation with multiple multi-antecedent rules, in *Proceedings of IEEE International Conference on Fuzzy Systems* (2012), pp. 1170–1177
97. S. Jin, R. Diao, C. Quek, Q. Shen, Backward fuzzy rule interpolation with multiple missing values, in *Proceedings of IEEE International Conference on Fuzzy Systems* (2013), pp. 1–8
98. S. Jin, R. Diao, C. Quek, Q. Shen, Backward fuzzy rule interpolation. IEEE Trans. Fuzzy Syst. **22**(6), 1682–1698 (2014)
99. S. Jin, R. Diao, C. Quek, Q. Shen, a-cut-based backward fuzzy interpolation, in *Proceedings of IEEE International Conference on Cognitive Informatics and Cognitive Computing* (2014), pp. 211–218
100. S. Jin, R. Diao, C. Quek, Q. Shen, Terrorism risk assessment using bidirectional hierarchical fuzzy rule interpolation. Under Rev. Potential J Publ. (2014)

Chapter 2
Background: Fuzzy Rule Interpolation

Conventional fuzzy reasoning methods such as Mamdani [1] and TSK [2, 3] require that the rule bases are dense. That is, the input universe of discourse is covered completely by the rule base. These methods determine output by rule matching, i.e. matching observed input to rule premises and calculating conclusions as weighted combinations of rule consequents with nonzero matching in which weights depend on the degree of matching. When an observation occurs, a consequence can always be derived by using such dense rule bases. However, due to incomplete knowledge, such as missing or insufficient expertise or available sample data for covering all possible input configurations, rule base construction may produce sparse or incomplete rule bases. In this case, the input universes of discourses may not be covered completely by the rule base(s), then conventional fuzzy reasoning methods can encounter difficulties if an observation occurs in a gap, resulting in no rule fired and thus no consequence derived.

The *sparse rule bases* problem was initially proposed in [4] and has also been termed "the tomato problem", shown in Fig. 2.1. Given an observation that a tomato is yellow and a rule base which contains two rules that "if a tomato is red, then it is ripe", and "if a tomato is green, then it is unripe", what is the conclusion? Intuitively, a consequence would be that the yellow tomato is half ripe. However, none of the conventional fuzzy inference approaches offers the ability to arrive at such a conclusion. To address this problem, fuzzy rule interpolation (FRI) was proposed in [5, 6].

When given observations have no overlap with antecedent values, no rule can be employed in classical fuzzy inference; therefore, no consequence can be derived. If this condition is violated, the rule base is considered to be sparse, i.e. it contains gaps. In sparse fuzzy rule bases, conventional fuzzy reasoning methods encounter difficulty because of the lack of inference evidence [5, 6]. Fuzzy rule-based interpolation techniques were introduced in order to allow inference in sparse fuzzy rule bases, and thus extend fuzzy inference mechanisms. Basically, FRI techniques perform interpolative approximate reasoning by taking into consideration the existing

S. Jin et al., *Backward Fuzzy Rule Interpolation*,
https://doi.org/10.1007/978-981-13-1654-8_2

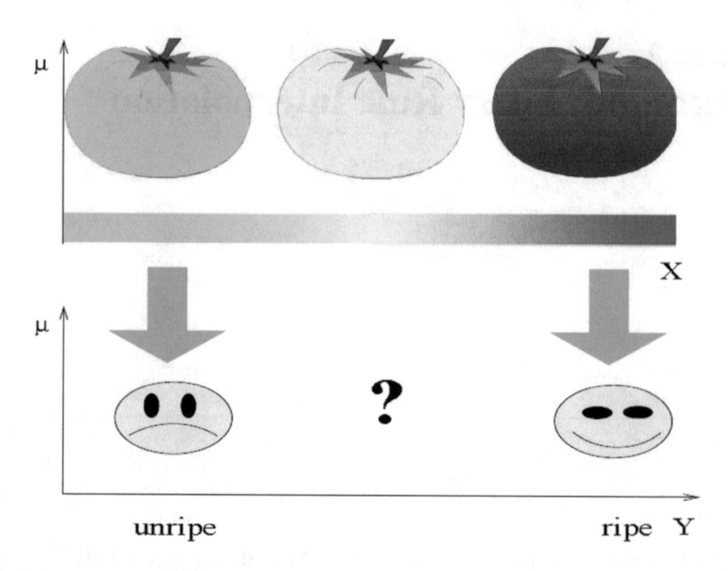

Fig. 2.1 Fuzzy reasoning of tomato problem

fuzzy rules for cases where there is no matching fuzzy rule. Interpolative reasoning, through a sparse rule base, may still obtain certain conclusions and thus improve the applicability of fuzzy models. Also, with the help of fuzzy interpolation, the complexity of a rule base can be reduced by omitting those fuzzy rules which may be approximated from their neighbouring ones [7, 8].

Successful applications of fuzzy interpolation have been reported in the literature [9, 10], including fuzzy control and fuzzy modelling. In particular, fuzzy interpolation has been used for the simulation of automated guided vehicles which are able to automatically track a given path but avoid any collision while on the move [11]. Fuzzy interpolation has also been used to support behaviour-based control [12] and user adaptive systems [13]. In order to perform prediction, proper modelling of the functional relationship between the input and output of the known data set is vital. Fuzzy interpolation has been applied to support fuzzy modelling in a broad range of application areas, particularly if only limited data are available [8, 14–17].

2.1 General Properties of FRI Methods

In this section, for the evaluation criterion of the different techniques based on the same fundamentals, the general properties related to fuzzy rule interpolative methods are briefly reviewed. The properties reflect an application-oriented viewpoint [18–22].

1. **Preservation of Convexity and Normality (CNF)**
   A FRI method should maintain the convexity and normality for any interpolative conclusion. That is, if an observation is convex and normal, then the interpolated result must also be convex and normal. The convexity condition dictates that membership function values must increase or decrease *monotonically* on either side of the maximum point:

$$\mu(\lambda x_1 + (1-\lambda)x_2) \geq min(\mu_A(x_1); \mu_A(x_2))$$

   where $\lambda \in [0, 1]$, $x_1, x_2 \in X$.
   The normality condition dictates that the resultant membership values produced by a fuzzy rule interpolation method should be in the range of [0, 1] only and should not produce more than one membership function value for one input.

2. **Applicability to Arbitrary Fuzzy Membership Functions (AAFMF)**
   A FRI method should ideally be able to deal with different kinds of membership function with different rules. For each $A^* \in X$ with a valid fuzzy membership function, the interpolated conclusion $B^* = I(A^*) \in Y$, where I is the interpolation process, should also be a valid fuzzy set for an arbitrary set of rules.
   This condition can be weakened practically to the case of piecewise linear and Gauss-bell-shaped fuzzy sets, being frequently encountered in the applications, and complicated, irregularly shaped input sets raise computational overhead.

3. **Mapping Similarity (MS)**
   Let $s_Z : \underline{Z} \times \underline{Z} \rightarrow \mathbb{U}$ denote the similarity function defined in the fuzzy sets of Z. Then, for $A^*, A_i^j \in X$, if $s_X(A^*, A_i^1) \geq s_X(A^*, A_i^2)$, $(i = 1, \cdots, N)$, then $s_Y(I(A_i^*), B^1) \geq s_Y(I(A_i^*), B^2)$, where $R_j : A_i^j \rightarrow B^j (j = 1, 2)$ are two rules of rule base R.

   This property indicates that "the more similar the observation to an antecedent, the more similar the conclusion should be to the corresponding consequent of the given antecedent".

4. **Preserving "In Between" (PIB)**
   In a linear interpolation, if $A_i^1 < A_i^* < A_i^2$, $(i = 1, \cdots, N)$, then $B^1 < I(A_i^*) < B^2$, where $R_j : A_i^j \rightarrow B^j (j = 1, 2)$ are two rules of rule base R. This property states that if the antecedent sets of two neighbouring rules "surround" an observation, the approximated conclusion should also "be surrounded" by the consequent sets of those rules.

5. **Compatibility with Rule Base (CRB)**
   Let all fuzzy sets in rules and the observation A^* be of the same type T. Mapping I is shape invariant, and $I(A^*)$ is also type T. This condition requires the validity of the *modus ponens*, i.e. if an observation coincides with the antecedent part of a rule, the conclusion produced by the method should correspond to the consequent part of that rule.

6. **Fuzziness of Approximated Result (FAR)**
   The less uncertain the observation is, the less fuzziness should be associated with approximated consequent. In other words, in the case of a singleton observation the method should produce a singleton consequence. Thus, a crisp conclusion can be expected only if all the consequents of the rules taken into consideration in the interpolation are singletons, i.e. the knowledge base produces certain information from fuzzy input data.
7. **Approximation Capability (AC)**
   The estimated rule should approximate with the highest possible degree the relationship between universes of the antecedent and consequent. If the number of the measurement points tends to infinity, the result should converge to the approximated function independent of the absolute position of the measurement points.
8. **Preservation of Piecewise Linearity (PPWL)**
   If the fuzzy sets of the rules taken into consideration are piecewise linear, the approximated sets should preserve this property. That is, interpolation can be computed using only characteristic points which describe a given polygonal fuzzy set, thereby ignoring any non-characteristic points and reducing computational effort.

   Actually, this condition is usually very difficult to enforce theoretically, and for a majority of the existing approaches, this does not strictly hold. However, in practice, piecewise linearity only means that non-characteristic points can be easily approximated to a certain accuracy (which most of the existing approaches do follow).
9. **Multiple Multi-antecedent Rules for Support (MMARS)**
   This condition indicates that an FRI technique should present similar characteristics when being extended and applied to multidimensional input spaces. This means that there should not be any restriction on the number of rules in the sparse rule base system although it may have a limited number of scattered rules.

   The original motivation behind FRI techniques was to reduce complexity, this is meaningful only in the case of many input dimensions, so FRI working only with a single-antecedent rule base has limited applicability. Surprisingly, researchers often neglect discussing this case, as pointed out in [16], although, it is not always straightforward to extend a single-antecedent method to the multi-antecedent case.

2.2 Categories of FRI Methods

In reviewing the development of fuzzy interpolation techniques, most can be categorised into one of two classes: single-step fuzzy rule interpolation and intermediate rule-based fuzzy interpolation, with some exceptions (e.g. type-2 fuzzy interpolation [23, 24]).

1. **Single-Step Fuzzy Rule Interpolation**
   The single-step rule interpolation category of approaches directly interpolates rules whose antecedents match the given observation. The consequence of the interpolated rule is thus the logical outcome. Typical approaches in this group [5, 6, 25] are based on the use of α-cut ($\alpha \in (0, 1]$). The α-cut of the interpolated consequent fuzzy set is calculated from the α-cuts of the observed antecedent fuzzy sets, and those of all the fuzzy sets involved in the rules used for interpolation. Having established the consequent α-cut values for all $\alpha \in (0, 1]$, the consequent fuzzy set is then assembled by applying the Resolution Principle [26–28]. Resolution Principle is a method of theorem proving that proceeds by constructing refutation proofs. Resolution Principle applies to first-order logic formulas in Skolemised form. These formulas are basically sets of clauses each of which is a disjunction of literals. Unification is a key technique in proofs by resolution.

2. **Intermediate Rule-Based Fuzzy Interpolation**
   The intermediate rule-based interpolation category is based on the analogical reasoning mechanism [19]. Such approaches first interpolate an artificially created intermediate rule such that the antecedents of the intermediate rule are similar to the given observation [18]. Then, a conclusion can be derived by firing this intermediate rule through analogical reasoning. The shape distinguishability between the resulting fuzzy set and the consequence of the intermediate rule is then analogous to the shape distinguishability between the observation and the antecedent of the created intermediate rule. Because this group of approaches always creates intermediate rules in the first instance, they are also called "intermediate rule-based fuzzy interpolation". In particular, the scale and move transformation-based approach (T-FRI) [7, 8], which belongs to this group, offers a flexible means to handle both interpolation and extrapolation involving multiple multi-antecedent rules. Therefore, the T-FRI method has been adopted as the foundation for the work in this book.

2.3 Single-Step Fuzzy Rule Interpolation Methods

2.3.1 KH Interpolation Method

The concept of α-cut distance-based KH (Kóczy–Hirota) fuzzy rule interpolation, is originally for fuzzy rule base complexity reduction [5, 6]. The starting point relies upon the Extension Principle (EP) [29, 30] and Resolution Principle (RP) [26–28]. EP extends a conventional mapping function $f : A_1 \times \cdots \times A_n \rightarrow B$, where $A_1, \ldots, A_n$ and B are crisp domains, to a fuzzy mapping f^*: $\mu_{A_1}(x_1) \times \cdots \times \mu_{A_n}(x_n) \rightarrow \mu_B(y)$ where, for every element x_i, $\mu_{A_i}(x_i)$ denotes the membership of x_i with respect to (fuzzy) set A_i. Formally, the formula to define f^* is: $\mu_B(y) = sup_{x_1,\ldots,x_n \rightarrow y} min\{\mu_{A_1}(x_1), \ldots, \mu_{A_n}(x_n)\}$, for all $x_1 \in A_1, \ldots, x_n \in A_n$

and $y \in B$. RE states that the solution of a problem for fuzzy sets can be found in the form of solving first for arbitrary α-cuts and then extending the solution to the fuzzy case. RP describes the decomposition of fuzzy sets to α-cuts:

$$A = \bigcup_{\alpha \in [0,1]} \alpha A_\alpha \tag{2.1}$$

where $\bigcup$ means maximum.

2.3.1.1 KH Interpolation with Two Single-Antecedent and Single-Consequent Rules

The basic idea behind any fuzzy interpolation method is to obtain a fuzzy conclusion if two rules and the observation are given (see Fig. 2.2). An important notion in interpolative reasoning is the "less than" relation between two continuous, valid and normal fuzzy sets. Fuzzy set A_1 is said to be less than A_2, denoted by $A_1 \prec A_2$, if $\forall \alpha \in [0, 1]$, the following conditions hold:

$$inf\{A_{1\alpha}\} < inf\{A_{2\alpha}\},\ sup\{A_{1\alpha}\} < sup\{A_{2\alpha}\} \tag{2.2}$$

where $A_{1\alpha}$ and $A_{2\alpha}$ are the α-cut of A_1 and that of A_2 respectively, $inf\{A_{i\alpha}\}$ is the infimum of $A_{i\alpha}$, and $sup\{A_{i\alpha}\}$ is the supremum of $A_{i\alpha}$, $i = 1, 2$.

For simplicity, suppose that the two following fuzzy rules are given:

$$\text{If } X \text{ is } A_1 \text{ then } Y \text{ is } B_1,$$
$$\text{If } X \text{ is } A_2 \text{ then } Y \text{ is } B_2,$$

Also, suppose that these two rules are adjacent or neighbouring, i.e. no rule exists such that the antecedent value A of that rule lies between the region of A_1 and A_2. To interpolate in the region between the antecedent values of these two rules, i.e. to determine a new conclusion B^* for an observation A^* located between fuzzy sets A_1 and A_2, rules in a given rule base are arranged with respect to a partial ordering amongst the valid and normal fuzzy sets of the antecedents variables. For the above two rules, this means that

$$A_1 \prec A^* \prec A_2 \tag{2.3}$$

In order to determine the fuzzy result B^*, KH interpolation uses the interpolation equation

$$\frac{d(A^*, A_1)}{d(A^*, A_2)} = \frac{d(B^*, B_1)}{d(B^*, B_2)} \tag{2.4}$$

where $d(.,.)$ is typically the Euclidean distance between two fuzzy sets (though other distance metrics may equally be used as alternatives). This is illustrated in Fig. 2.2,

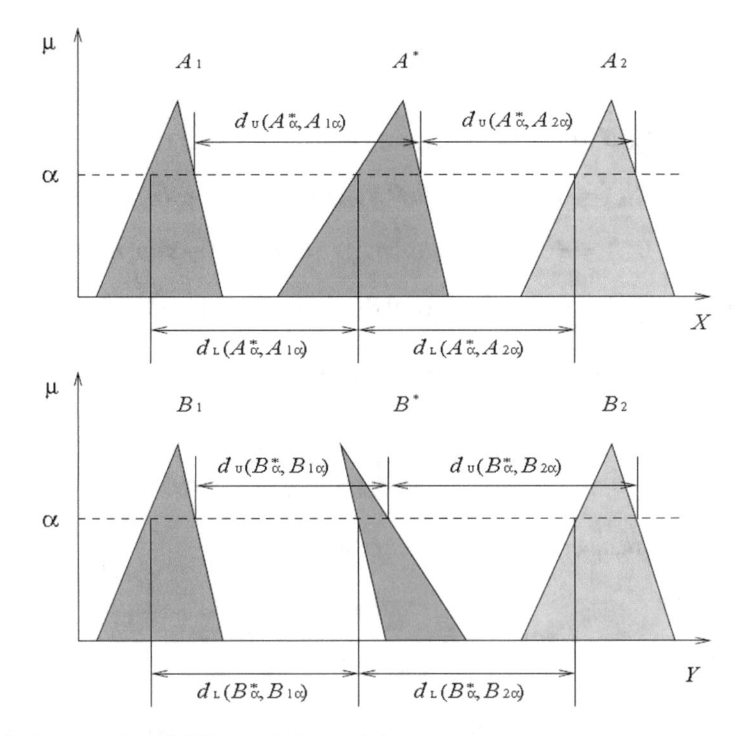

Fig. 2.2 An example of KH fuzzy rule interpolation

where the lower and upper distances between α-cuts, $A_{1\alpha}$ and $A_{2\alpha}$ are defined as follows:

$$d_L(A_{1\alpha}, A_{2\alpha}) = d_L(A^*_\alpha, A_{1\alpha}) + d_L(A^*_\alpha, A_{2\alpha}) = d(inf\{A_{1\alpha}\}, inf\{A_{2\alpha}\}) \quad (2.5)$$

$$d_U(A_{1\alpha}, A_{2\alpha}) = d_U(A^*_\alpha, A_{1\alpha}) + d_U(A^*_\alpha, A_{2\alpha}) = d(sup\{A_{1\alpha}\}, sup\{A_{2\alpha}\}) \quad (2.6)$$

From Eq. 2.4, the α-cuts interpolation equations can be obtained as:

$$\frac{d(A^*_\alpha, A_{1\alpha})}{d(A^*_\alpha, A_{2\alpha})} = \frac{d(B^*_\alpha, B_{1\alpha})}{d(B^*_\alpha, B_{2\alpha})} \quad (2.7)$$

$$\frac{d(A^*_\alpha, A_{1\alpha})}{d(A^*_\alpha, A_{2\alpha})} = \frac{d(B^*_\alpha, B_{1\alpha})}{d(B^*_\alpha, B_{2\alpha})} \quad (2.8)$$

From Eqs. 2.5 and 2.6, the lower and upper distances between α-cuts, $B_{1\alpha}$, $B_{2\alpha}$ and B^*_α are defined as follows:

$$d_L(B^*_\alpha, B_{1\alpha}) = d(inf\{B^*_\alpha\}, inf\{B_{1\alpha}\}) = inf\{B^*_\alpha\} - inf\{B_{1\alpha}\} \quad (2.9)$$

$$d_L(B_{2\alpha}, B_{1\alpha}) = d(inf\{B_{2\alpha}\}, inf\{B_{1\alpha}\}) = inf\{B_\alpha\} - inf\{B_{1\alpha}\} \quad (2.10)$$

$$d_U(B^*_\alpha, B_{1\alpha}) = d(sup\{B^*_\alpha\}, sup\{B_{1\alpha}\}) = sup\{B^*_\alpha\} - sup\{B_{1\alpha}\} \quad (2.11)$$

$$d_U(B_{2\alpha}, B_{1\alpha}) = d(sup\{B_{2\alpha}\}, sup\{B_{1\alpha}\}) = sup\{B_\alpha\} - sup\{B_{1\alpha}\} \quad (2.12)$$

Applying these results, Eq. 2.7 can be rewritten as:

$$inf\{B^*_\alpha\} = \frac{\frac{inf\{B_{1\alpha}\}}{d_L(A^*_\alpha, A_{1\alpha})} + \frac{inf\{B_{2\alpha}\}}{d_L(A^*_\alpha, A_{2\alpha})}}{\frac{1}{d_L(A^*_\alpha, A_{1\alpha})} + \frac{1}{d_L(A^*_\alpha, A_{2\alpha})}} \quad (2.13)$$

and in the same manner, Eq. 2.8 can be rewritten as

$$sup\{B^*_\alpha\} = \frac{\frac{sup\{B_{1\alpha}\}}{d_U(A^*_\alpha, A_{1\alpha})} + \frac{sup\{B_{2\alpha}\}}{d_U(A^*_\alpha, A_{2\alpha})}}{\frac{1}{d_U(A^*_\alpha, A_{1\alpha})} + \frac{1}{d_U(A^*_\alpha, A_{2\alpha})}} \quad (2.14)$$

From this, $B^*_\alpha = (inf\{B^*_\alpha\}, sup\{B^*_\alpha\})$ results. And the conclusion fuzzy set B^* can be constructed using the representation principle of fuzzy sets:

$$B^* = \bigcup_{\alpha \in [0,1]} \alpha B^*_\alpha \quad (2.15)$$

Despite this method's capability of handling the tomato problem, it does not guarantee validity, although they may be normal, as B^* shows in Fig. 2.2.

2.3.1.2 KH Interpolation for Multiple Multi-antecedent Rules

Without loss of generality, given an observation O and its N closest rules with M antecedents, they can be expressed in the following format:

$O : A^*_1, \cdots, A^*_k, \cdots, A^*_M$
R_1 : IFx_1 is $A^1_1, \cdots$, and x_k is $A^1_k, \cdots$, and x_M isA^1_M, THEN y is B^1
R_2 : IF x_1 is $A^2_1, \cdots$, and x_k is $A^2_k, \cdots$, and x_M isA^2_M, THEN y is B^2
......
R_N : IF x_1 is $A^N_1, \cdots$, and x_k is $A^N_k, \cdots$, and x_M isA^N_M, THEN y is B^N

In order to enable interpolation with multiple rules, firstly, the distances in different dimensions need to be aggregated in order to express the overall "distance" between the given observation and the N rules used for interpolation. However, the absolute distances within different dimensions may not be compatible because different attributes have different domains. To make these compatible, the normalised attribute distance is defined by:

$$d'_L(A^*_{k\alpha}, A^n_{k\alpha}) = \frac{d(inf\{A^*_{k\alpha}\}, inf\{A^n_{k\alpha}\})}{sup_k - inf_k} \tag{2.16}$$

$$d'_U(A^*_{k\alpha}, A^n_{k\alpha}) = \frac{d(sup\{A^*_{k\alpha}\}, sup\{A^n_{k\alpha}\})}{sup_k - inf_k} \tag{2.17}$$

where sup_k and inf_k are the supremum and infimum in the domain of variable x_k. Then, the overall distance between the observation O and the antecedents of rule R_i, denoted as $D_L(O; Ri)$ and $D_U(O; Ri)$, is calculated by:

$$D_L(O, Ri) = \sqrt{\sum_{k=1}^{M}(d'_L(A^*_{k\alpha}, A^n_{k\alpha}))^2} \tag{2.18}$$

$$D_U(O, Ri) = \sqrt{\sum_{k=1}^{M}(d'_U(A^*_{k\alpha}, A^n_{k\alpha}))^2} \tag{2.19}$$

From this, the remainder of the calculation is the same as that of single-antecedent rule interpolation, introduced in the last section. In order to enable interpolation with multiple rules, the weighting scheme needs to be generalised. Again, without loss of generality, the weight of R_i can be calculated as follows:

$$(\omega^i_\alpha)_L = \frac{\frac{1}{D_L(O,Ri)}}{\Sigma_i^N \frac{1}{D_L(O,Ri)}} \tag{2.20}$$

$$(\omega^i_\alpha)_U = \frac{\frac{1}{D_U(O,Ri)}}{\Sigma_i^N \frac{1}{D_U(O,Ri)}} \tag{2.21}$$

$inf\{B^i_\alpha\}$ and $sup\{B^i_\alpha\}$ can be calculated according to Eqs. 2.13 and 2.14. The consequence B^*_α can then be assembled:

$$inf\{B^*_\alpha\} = \Sigma_i^N(\omega^i_\alpha)_L inf\{B^i_\alpha\} \tag{2.22}$$

$$sup\{B^*_\alpha\} = \Sigma_i^N(\omega^i_\alpha)_U sup\{B^i_\alpha\} \tag{2.23}$$

This technique has many advantages as it behaves approximately linearly between the α levels. It is more suitable for triangular- and trapezoidal-shaped fuzzy sets because these can be easily described by few characteristic points that the α-cut process. However, it is restricted to convex and normal fuzzy sets. This method also does not always provide a normal conclusion and or maintain piecewise linearity of the resultant conclusion. Sometimes the bounds of the results are not in the expected order, because the interpolation weights for the left-hand and right-hand sides are not related to one another.

2.3.2 *Other Single-Step Fuzzy Rule Interpolation Methods*

In order to address the possibility that KH interpolation may return an abnormal result, as shown in Fig. 2.2, a number of modifications or improvements of the original approach have been proposed in the literature. The HCL (Hsiao, Chen and Lee) approach [31] eliminates this problem by fixing the core of the consequence generated by the KH approach and shifting its support along with the consequent variable axis. The HCL interpolation method is an interpolative reasoning method based on the KH method. The difference is that it not only interpolates the "bottoms" of the fuzzy sets, but also interpolates the highest points. It can guarantee that "If fuzzy rules $A_1 \Rightarrow B_1$, $A_2 \Rightarrow B_2$ and the observation A^* are defined by triangular membership functions, the interpolated conclusion B^* will also be a triangular-type". However, this method is specifically designed for triangular cases, and thus the piecewise linearity property is not preserved for trapezoidal fuzzy sets or others.

Another important modification to the original KH approach represents each fuzzy set by using two vectors of characteristic points, which describe the left and right slopes [32–39]. By noting one of the conditions that ensure the validity of the consequent fuzzy set, the approach interpolates the consequence of a given observation by means of transformation of the two vectors. It does this in an effort to satisfy the discovered condition and thus ensure that the interpolated result is a normal and convex fuzzy set. This approach has been improved by using fuzziness conservation techniques [16, 40]. In this case, the fuzziness of the consequence is determined by both the fuzziness of the given observation and the fuzziness of the fuzzy sets involved in the rules used for interpolation.

2.3.2.1 The HCL Interpolation Method

The HCL interpolation method calculates the bottom of B^* in the same way as the KH method does, but calculates the top point in a different way. Figure 2.3 illustrates a typical fuzzy interpolation problem, where $k_1, t_1, k, t, k_2, t_2, h_1, m_1, h, m, h_2$ and m_2 represent the slopes of corresponding fuzzy sets. The process to determine the top point of B^* can be described as follows:

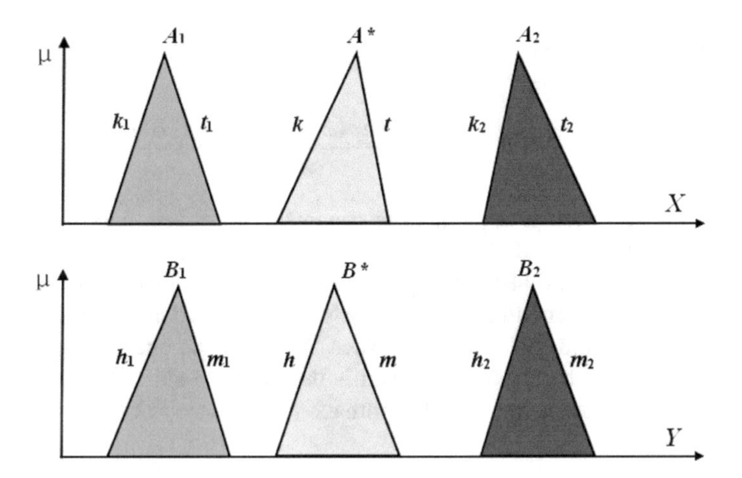

Fig. 2.3 HCL fuzzy rule interpolation

1. Deciding the slopes h and m of the triangular-type membership function B^*. Let:

$$k = k_1 x + k_2 y \tag{2.24}$$

$$t = t_1 x + t_2 y \tag{2.25}$$

where x and y are real numbers. If $\frac{k_1}{t_1} \neq \frac{k_2}{t_2}$, then unique x and y are computed by solving Eqs. 2.24 and 2.25 simultaneously. Let

$$h = |h_1 x + h_2 y| c \tag{2.26}$$

$$m = -|m_1 x + m_2 y| c \tag{2.27}$$

where c is a constant. Otherwise, let

$$h = kc \tag{2.28}$$

$$m = tc \tag{2.29}$$

again (where c is a constant).

2. The position of the top point b_1^* is calculated by solving the following equation,

$$\frac{1}{CP(B^*) - inf(B^*)} : \frac{-1}{sup(B^*) - CP(B^*)} = h : m \tag{2.30}$$

where $CP(A)$ is the centre point of the specified fuzzy set A. $CP(A) = (\underline{A_\alpha} + \overline{A_\alpha})/2$, where $\alpha = height(A)$. A_α denotes the α-cut of A. The centre point of a triangular fuzzy set is just its top point (of membership value of 1). Equation 2.30 can be reformulated as

$$CP(B^*) = \frac{m \times sup(B^*) - h \times inf(B^*)}{m - h} \tag{2.31}$$

Note that this approach is applicable to triangular fuzzy sets only. In particular, it represents both slopes of each fuzzy set as a linear function. The slopes of the consequent fuzzy set are also linear functions whose parameters are interpolated from those of the observation and the fuzzy sets involved in the rules are used for interpolation. From this, a ratio between the left slope and the right slope of the consequence can be calculated which is then used to shift the support of the generated consequence using the KH approach in reference to the normal point of the consequence.

2.3.2.2 The FIVE Interpolation Method

Vague environment-based interpolation method (FIVE) [41] is based on the concept of similarity or indistinguishability of the elements. The equality relation $\approx$ is a fundamental concept in the FIVE algorithm, which is also called a similarity relation or indistinguishability operator. It is defined on the set X as a mapping $E : X \times X \to [0, 1]$ that satisfies the following three axioms:

$$\begin{aligned} E_{\approx}(x, x) &= 1; \quad \text{(reflexivity)} \\ E_{\approx}(x, y) &= E_{\approx}(y, x); \quad \text{(symmetry)} \\ T(E_{\approx}(x, y), E_{\approx}(y, z)) &\leq T(E_{\approx}(x, z)) \quad \text{(monotonicity)} \end{aligned} \tag{2.32}$$

where X is the underlying domain and T is any lower semi-continuous t-norm [42].

The concept of vague environment is based on the similarity or indistinguishability of the elements. Two values in the vague environment are ε-distinguishable if their distance is grater then ε. The distances in vague environment are weighted distances. Two values in the vague environment X are ε-distinguishable if

$$\varepsilon > \delta_s(x_1, x_2) = \left| \int_{x_2}^{x_1} s(x)dx \right| \tag{2.33}$$

where $\delta_s(x_1, x_2)$ is the vague distance of the values x_1, x_2 and $s(x)$ is the scaling function on X.

If the vague environment of a fuzzy partition (the scaling function or the approximative scaling function) exists, the member sets of the fuzzy partition can be characterised by points in the vague environment. If all the vague environments of the antecedent and consequent universes of the fuzzy rule base exist, all of the primary fuzzy sets (linguistic terms) used in the fuzzy rule base can be characterised by points in their vague environment. So, the fuzzy rules (built on the primary fuzzy sets) can be characterised by points in the vague environment of the fuzzy rule base too. In this case, the approximative fuzzy reasoning can be handled as a classical interpolation task. The FIVE algorithm is implemented with the following steps.

1. Determine the similarity of two fuzzy sets and the vague distance of points in a vague environment.
2. Generate a vague environment from the fuzzy partition of the linguistic terms within the fuzzy rules.
3. Decide on the approximate scaling function.
4. Calculate the conclusion by approximating the vague points of the rule base using any classical interpolation method.

For using of this method in a single-dimensional antecedent universe case, the linear rule interpolation of two fuzzy rules for a vague environment is adopted in this method. Note that the restrictions for the two chosen rules are the flanking of the observation on the antecedent side and the existence of the ordering on the consequent side. Without loss of generality, the consequence universe of the fuzzy rules is one dimensional (the multidimensional case can be decomposed to a single-dimensional one) and the antecedent universe is multidimensional. The process of the example is outlined in the pseudocode shown in Algorithm 2.3.1.

Algorithm 2.3.1: FIVE Interpolation (X,Y, α, ϵ)

X: the set of input universe;
Y: the set of output universe;
A_1, A_2: the antecedent fuzzy sets;
B_1, B_2: the consequent fuzzy sets;
A^*: the observation fuzzy sets;
B^*: the conclusion fuzzy sets;
μ: the fuzzy membership function;
α: the α-cut level;
ϵ: the indistinguishability parameter;
δ_s: the vague distance;
S: the scaling function.

$A_1, A_2 \in X, B_1, B_2 \in Y$
$A_1 \prec A^* \prec A_2, B_1 \prec B_2$
$\forall \alpha \in (0, 1]$
$\delta_s(x_1, x_2) \leftarrow |\int_{x_2}^{x_1} S(x)dx|$
$\mu_A(x) \leftarrow 1 - min\{\delta_s(a, b), 1\}$
$\mu_A(x) \leftarrow 1 - min\{|\int_a^b S(x)dx|, 1\}$
$\delta_s(a, b) \leftarrow 1 - \alpha$
$S(x) = |\mu'(x)| \leftarrow |d\mu/dx|$
$min\{\mu_i(x), \mu_j(x)\} > 0 \Rightarrow |\mu_j(x)| = |\mu_j(x)|, \forall i, j \in I$
$S(x) \leftarrow Approximate[S(x)]$
$B^* \leftarrow ClassicalInterpolation(A_1, A_2, A^*, B_1, B_2)$

The FIVE method is basically developed as an alternative to the compositional rule of inference (CRI) [41]. The crisp conclusion obtained using this method has a number of advantages over that achieved using classical min-max CRI. In particular, the control function of this method always fits the points of the fuzzy rules, while the control function of the CRI does not. This method can be used for either dense rule bases or for sparse rule bases. A crisp conclusion is obtained directly from the vague conclusion without any defuzzification. If a fuzzy conclusion is required, the crisp point which is produced by this method should be converted to a fuzzy value. "FIVE" is restricted to convex and normal fuzzy sets. In general, it does provide a normal conclusion but this may depend on the proper approximate scaling function. It also supports multiple multi-antecedent rules.

The accuracy of this method depends upon the approximate scaling function and how it is determined. This function is critical for each problem dimension as it represents the vague environment for a given problem. Once the vague environments for the antecedent and consequent universes are determined, any rule can be represented by a single point in the environments and then linear interpolation can be used to calculate the conclusion. However, such environments for both the antecedent and the consequent are required in advance.

2.3.2.3 The MACI Method

The modified α-cut-based fuzzy interpolation (MACI) technique [34] eliminates the normality problem, while simultaneously attempting to maintain the advantageous computational properties of the original KH method.

- **MACI with Triangular-Shape Fuzzy Sets in Single-Dimensional Input Space** Let the fuzzy sets in both the input and output spaces have isosceles triangular shape. Then, a fuzzy set A can be described by its characteristic points as $A = \{a_{-1}, a_0, a_1\}$, where a_{-1} and a_1 denote the infimum and the supremum of the support of fuzzy set A, respectively, and a_0 denotes the single core of A. $a_{-1} \leq a_0 \leq a_1$ should be satisfied to ensure the normality of A. Suppose that vectors α contain the characteristic values of the right flank of A as

$$\alpha = [a_0 \; a_1]^T \tag{2.34}$$

The conclusion according to the original α-cut-based KH fuzzy rule interpolation method by means of the vector representation form can be expressed (for the right flank) as:

$$b^* = (I - I\Lambda)b_1 + I\Lambda b_2 \tag{2.35}$$

where I is the identity matrix,

$$\Lambda = [\lambda_0 \; \lambda_1] \tag{2.36}$$

Fig. 2.4 The definitions of input and output space

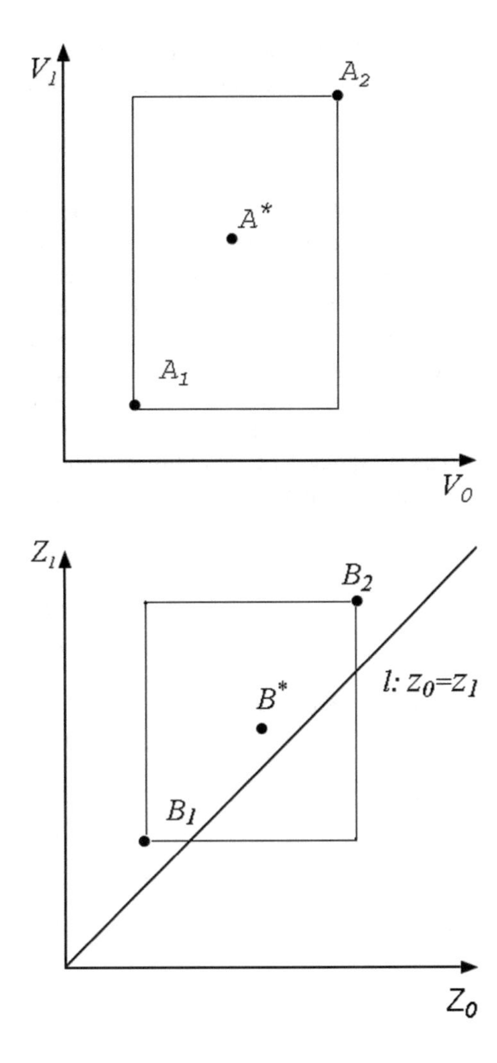

and

$$\lambda_k = \frac{a_k^* - a_{1k}}{a_{2k} - a_{1k}}, \quad k = 0, 1 \tag{2.37}$$

Two-dimensional space $V_0 \times V_1$ is defined in Fig. 2.4 for the first and the second elements of vectors a_1, a_2 and a^* as the right flank of fuzzy sets A_1, A_2 and A^*, respectively. Because by the premises of the original KH fuzzy rule interpolation method $a_{1k} < a_k^* < a_{2k}$, the fractions $\lambda_k (k = 0, 1)$ are non-negative number in the interval [0, 1]. Point B^* is obtained in the rectangle drawn by thin lines in

Fig. 2.4. In order to ensure that $inf\{B^*_{\alpha_1}\} \leq inf\{B^*_{\alpha_2}\} \leq sup\{B^*_{\alpha_2}\} \leq sup\{B^*_{\alpha_1}\}$ the coordinates of the conclusion B^*? should satisfy

$$b_0^* \leq b_1^* \tag{2.38}$$

That is, it should be above the straight line l: $z_0 = z_1$. If the rectangular is crossed by the line l, then there is always a chance of a non-normal conclusion. The whole rectangle is above the line l if the sets B_1 and B_2 overlap.
Transforming the points B_1, B_2 in another coordinate system where axis Z_0 is substituted by the straight line l: $z_0 = z_1$ and axis Z_1 remains unchanged. The coordinates of the conclusion are computed in the transformed coordinate system which ensures that it will never be under the straight line l. Then the resulting conclusion is transformed back to the original coordinate system.
The coordinates of an arbitrary vector b representing a CNF set from the output space can be calculated in the new system, thus:

$$\begin{aligned} b &= [b_0\, b_1] \rightarrow b' = [b_0'\, b_1'] \\ b_0' &= b_0 \cdot \sqrt{2} \\ b_1' &= b_1 - b_0 \end{aligned} \tag{2.39}$$

In matrix description

$$b' = bT \tag{2.40}$$

where

$$T = \begin{bmatrix} \sqrt{2} & -1 \\ 0 & 1 \end{bmatrix} \tag{2.41}$$

Then the transformed conclusion can be obtained as

$$b_0^{*\prime} = (1 - \lambda_0)b_{10}' + \lambda_0 b_{20}' \tag{2.42}$$

$$b_1^{*\prime} = (1 - \lambda_1)b_{11}' + \lambda_1 b_{21}' \tag{2.43}$$

with $\lambda_k (k = 0, 1)$, Eq. 2.35 can be transformed in matrix form

$$b^{*\prime} = (I - I\Lambda)b_1' + I\Lambda b_2' \tag{2.44}$$

The conclusion then can be transformed back by means of the coordinate Eq. 2.39 as

$$\begin{aligned} b_0^* &= b_0^{*\prime}/\sqrt{2} \\ b_1^* &= b_1^{*\prime} + b_0^* = b_1^{*\prime} + b_0^{*\prime}/\sqrt{2} \end{aligned} \tag{2.45}$$

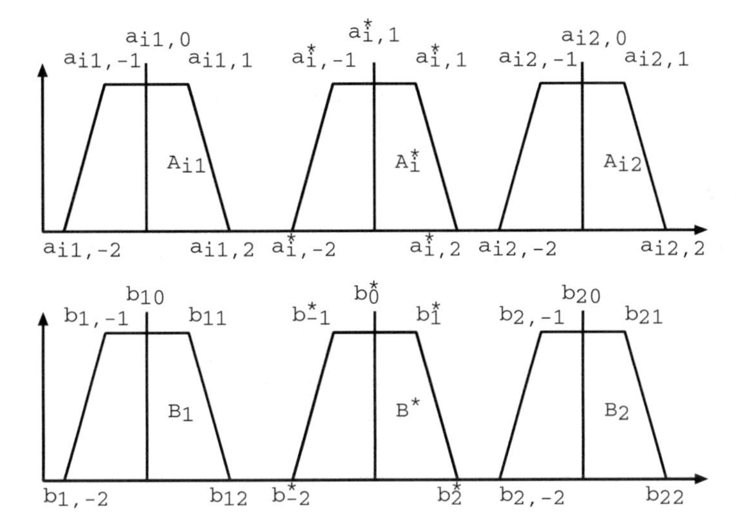

Fig. 2.5 Notations used for MACI method

or in matrix form

$$b^* = b^{*\prime}T^{-1} \tag{2.46}$$

where

$$T^{-1} = \begin{bmatrix} 1/\sqrt{2} & 1/\sqrt{2} \\ 0 & 1 \end{bmatrix} \tag{2.47}$$

- **MACI with Trapezoidal-Shape Fuzzy Sets in Multidimensional Input Spaces** For the sake of analysis, the notation for two trapezoidal rules shown in Fig. 2.5 are used. For the case where there are multidimensional input spaces with trapezoidal sets, the reference points for all membership functions can be calculated by taking the mid-point of the membership function. For K input dimensions, the reference point of the interpolated conclusion for trapezoidal membership function is:

$$b_0^* = (1 - \lambda_{core})b_{10} + \lambda_{core}b_{20} \tag{2.48}$$

where

$$\lambda_{core} = \frac{\sqrt{\Sigma_{i=1}^{k}(a_{i,0}^* - a_{i1,0})^2}}{\sqrt{\Sigma_{i=1}^{k}(a_{i2,0} - a_{i1,0})^2}} \tag{2.49}$$

Using the reference point, the right core can be calculated as:

$$b_1^* = (1 - \lambda_{right})b_{11} + \lambda_{right}b_{21} + (\lambda_{core} - \lambda_{right})(b_{20} - b_{10}) \tag{2.50}$$

where

$$\lambda_{right} = \frac{\sqrt{\Sigma_{i=1}^{k}(a_{i,1}^{*} - a_{i1,1})^2}}{\sqrt{\Sigma_{i=1}^{k}(a_{i2,1} - a_{i1,1})^2} \tag{2.51}$$

and the left core is:

$$b_{-1}^{*} = (1 - \lambda_{left})b_{1,-1} + \lambda_{left}b_{2,-1} + (\lambda_{core} - \lambda_{left})(b_{20} - b_{10}) \tag{2.52}$$

where

$$\lambda_{left} = \frac{\sqrt{\Sigma_{i=1}^{k}(a_{i,-1}^{*} - a_{i1,-1})^2}}{\sqrt{\Sigma_{i=1}^{k}(a_{i2,-1} - a_{i1,-1})^2} \tag{2.53}$$

For the right flank:

$$\begin{aligned} b_2^{*} = (1 - \lambda_{rightflk})b_{12} + \lambda_{right}b_{22} + (\lambda_{core} - \lambda_{right})(b_{20} - b_{10}) \\ + (\lambda_{right} - \lambda_{rightflk})(b_{21} - b_{11}) \end{aligned} \tag{2.54}$$

where

$$\lambda_{rightflk} = \frac{\sqrt{\Sigma_{i=1}^{k}(a_{i,2}^{*} - a_{i1,2})^2}}{\sqrt{\Sigma_{i=1}^{k}(a_{i2,2} - a_{i1,2})^2} \tag{2.55}$$

The left flank is calculated in a similar way.

There are many benefits related to this method, for example it is valid on the intervals between the characteristic points. Thus, it is sufficient to calculate only the characteristic points of the conclusion, which reduces the computational complexity. It also provides for better linear approximation behaviour between the breakpoints. This method preserves the fuzziness of the given rule base and supports multiple multi-antecedent rules. The original version of MACI was confined only to CNF sets, and its modified version can handle non-convex fuzzy sets as well as non-normal fuzzy sets. MACI will only yield a singleton conclusion if and only if the consequents are singletons themselves.

2.4 Intermediate Rule-Based Fuzzy Interpolation Methods

A number of ways of generating an intermediate rule which can then be used to infer a conclusion from the given observation have been developed in the literature. These include [7, 8, 18, 43–49] amongst others.

2.4.1 Scale and Move Transformation-Based FRI (T-FRI)

The scale and move transformation-based approach is able to handle interpolation with multiple antecedent rules, and the object value of each antecedent variable may be of triangular, complex polygonal, Gaussian or other bell-shaped fuzzy membership functions [7, 8, 44].

2.4.1.1 Properties and Base Process of T-FRI

T-FRI has the following properties:

1. It can handle both interpolation and extrapolation which involve multiple fuzzy rules, where each rule has multiple antecedents.
2. It guarantees the uniqueness as well as the normality and convexity of the resulting interpolated fuzzy sets.
3. It preserves piecewise linearity such that interpolation can be computed using only characteristic points which describe a given polygonal fuzzy set, thereby ignoring any non-characteristic points and saving computational effort.
4. It has been applied to problems such as truck backer-upper control and computer activity prediction [8].

The scale and move transformation-based interpolation procedure is illustrated in Fig. 2.6. Given an observation (input), it first uses real numbers termed: representative values, to represent the overall positions of the fuzzy sets of the observation and the original rule base within the domain of the variables, Then, the relative placement relation between the observation and the antecedents of the neighbouring rules for interpolation is obtained, termed: relative placement factor. An intermediate rule can be interpolated by applying the relative placement factor applied to both the antecedents and consequents of the neighbouring rules for interpolation. The representative value of the resulting antecedent is guaranteed to be equal to that of the observation, though the two fuzzy sets are usually non-identical. Next, the similarity degree between the observation and intermediate values is calculated using a predefined similarity measure. Specifically, scale rate s, scale ratio $\mathbb{S}$ and move rate m are used in scale and move transformation based interpolation to represent the similarity degrees. Finally, the consequence of the interpolated rule is computed by applying the transformation function to the consequent (output) while imposing the same similarity degree.

Trapezoidal and triangular membership functions are the fuzzy set representations, which are frequently used in fuzzy systems. Therefore, they are adopted in the algorithm description below. A key concept used in T-FRI is that of the representative value $Rep(A)$ for a given fuzzy set A. When trapezoidal representation is used, $Rep(A)$ is defined as the centre of gravity of its four points (a_0, a_1, a_2, a_3):

$$Rep(A) = \frac{a_0 + \frac{a_1+a_2}{2} + a_3}{3} \tag{2.56}$$

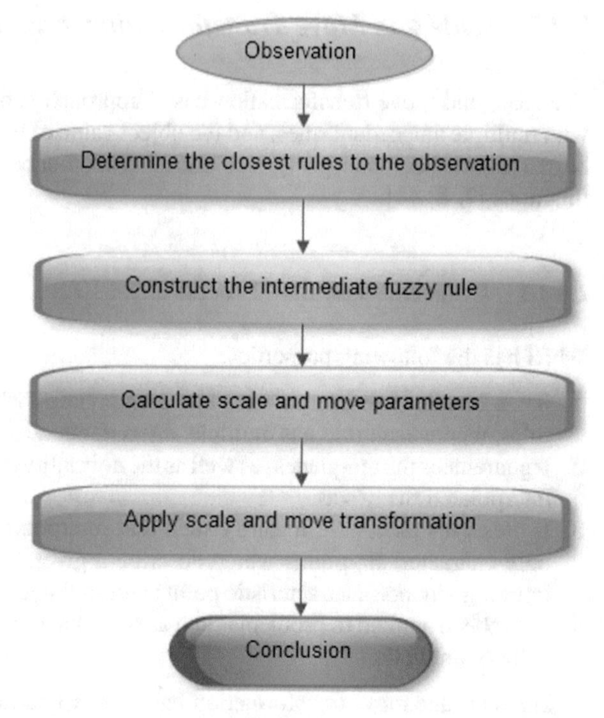

Fig. 2.6 Scale and move transformation-based interpolation

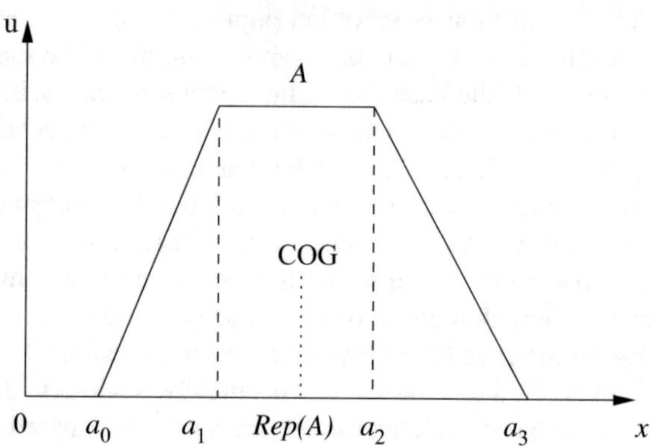

Fig. 2.7 The representative value for a trapezoidal fuzzy set

where a_0, a_3 represent the left and right extremities (with membership values 0), and a_1, a_2 denote the normal points (with membership value 1), as shown in Fig. 2.7.

As a specific case of trapezoids, where a_1 and a_2 are collapsed into a single value a_1, the fuzzy set becomes a triangular set (a_0, a_1, a_2). In particular, the corresponding $Rep(A)$ degenerates to the average value of the triple, as given below and shown in Fig. 2.8:

$$Rep(A) = \frac{a_0 + a_1 + a_2}{3} \tag{2.57}$$

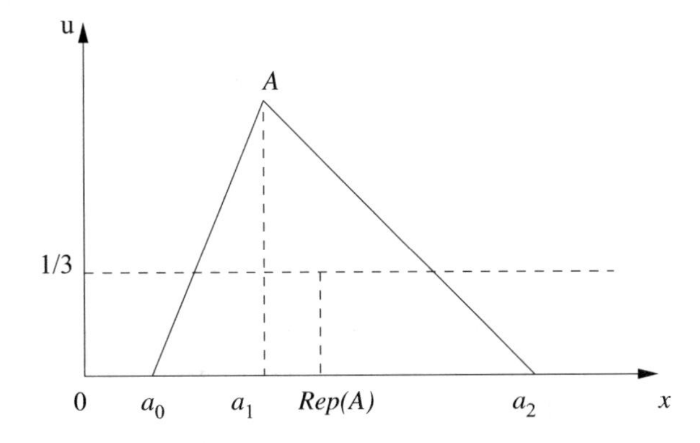

Fig. 2.8 The representative value for a triangular fuzzy set

In the following, the T-FRI method is outlined using trapezoidal fuzzy sets unless otherwise stated.

2.4.1.2 T-FRI with Two Triangular Single-Antecedent Rules

1. *Determination of Two Closest Rules*
   For single-antecedent rules $A \Rightarrow B$, the distances to the observation A^* can be computed using Eq. 2.58.

$$d = d(A, A^*) = d(Rep(A), Rep(A^*)) \tag{2.58}$$

2. *Construct Intermediate Fuzzy Terms*
   Suppose that the two neighbouring rules after distance comparison $A_1 \Rightarrow B_1$, $A_2 \Rightarrow B_2$, and the observation A^* are given as illustrated in Fig. 2.9. The intermediate fuzzy term $A^{'} = (1 - \lambda_A)A_1 + \lambda_A A_2$ can then be defined according to the ratio of distances λ_A between their representative values, and $Rep(A^{'}) = Rep(A^*)$:

$$\lambda_A = \frac{d(A_1, A^*)}{d(A_1, A_2)} = \frac{d(Rep(A_1), Rep(A^*))}{d(Rep(A_1), Rep(A_2))} \tag{2.59}$$

$$\begin{cases} {a_0}' = (1 - \lambda_A)a_{10} + \lambda_A a_{20} \\ {a_1}' = (1 - \lambda_A)a_{11} + \lambda_A a_{21} \\ {a_2}' = (1 - \lambda_A)a_{12} + \lambda_A a_{22} \end{cases} \tag{2.60}$$

   Similarly, the fuzzy set $B^{'}$ on the consequence domain can be obtained. In the single-antecedent case, $\lambda_B = \lambda_A$.

$$B^{'} = (1 - \lambda_B)B_1 + \lambda_B B_2 \tag{2.61}$$

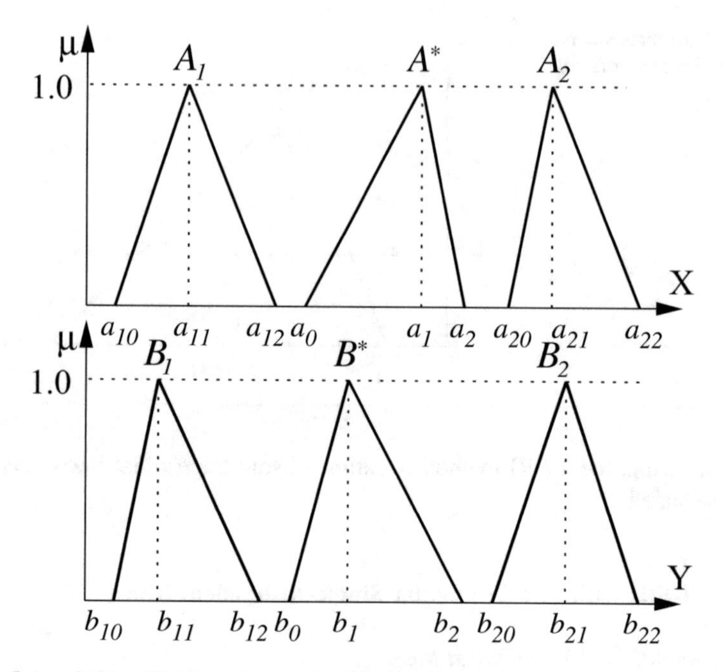

Fig. 2.9 Interpolation with triangular membership functions

3. *Scale Transformation*:
   Let $A'' = (a_0'', a_1'', a_2'')$ denote the fuzzy set generated by the scale transformation. By using the scale rate s_A, A''s current support (a_0', a_2') is transformed into a new support (a_0'', a_2''), such that $a_2'' - a_0'' = s_A \times (a_2' - a_0')$.

$$\begin{cases} a_0'' = \frac{a_0'(1+2s_A)+a_1'(1-s_A)+a_2'(1-s_A)}{3} \\ a_1'' = \frac{a_0'(1-s_A)+a_1'(1+2s_A)+a_2'(1-s_A)}{3} \\ a_2'' = \frac{a_0'(1-s_A)+a_1'(1-s_A)+a_2'(1+2s_A)}{3} \\ s_A = \frac{a_2''-a_0''}{a_2'-a_0'} \end{cases} \tag{2.62}$$

4. *Move Transformation*:
   The current support of A'' is moved to (a_0, a_2) while keeping its representative value, resulting in the fuzzy set A^*.

$$\begin{cases} m_A = \frac{a_0-a_0''}{\frac{a_1''-a_0''}{3}}, a_0 \geq a_0'' \\ m_A = \frac{a_0-a_0''}{\frac{a_2''-a_1''}{3}}, otherwise \end{cases} \tag{2.63}$$

Given the move ratio m_A, the transformed fuzzy set A^* can be calculated using:

$$\begin{cases} \begin{cases} a_0 = a_0'' + m_A \frac{a_1''-a_0''}{3} \\ a_1 = a_1'' - 2m_A \frac{a_1''-a_0''}{3} \\ a_2 = a_2'' + m_A \frac{a_1'-a_0''}{3} \end{cases} & m_A \geq 0 \quad (2.64a) \\ \begin{cases} a_0 = a_0'' + m_A \frac{a_2''-a_1''}{3} \\ a_1 = a_1'' - 2m_A \frac{a_2''-a_1''}{3} \\ a_2 = a_2'' + m_A \frac{a_2''-a_1''}{3} \end{cases} & otherwise \quad (2.64b) \end{cases}$$

5. The above transformations from $A^{'}$ to A^* can be concisely represented by the function $T(A^{'}, A^*)$. Similarly, the function T is applied to transforming $B^{'}$ to B^* such that:

$$T(B^{'}, B^*) = T(A^{'}, A^*) \quad (2.65)$$

where $s_B = s_A$ and $m_B = m_A$ for the current single-antecedent case.

2.4.1.3 T-FRI with Triangular Multiple Antecedent Variables

1. *Determination of Two Closest Rules*
   Without loss of generality, rules R_i, R_j and observation O can be represented in the following form:

$$\begin{array}{ll} R_i: & IF\ x_1 is A_1^i, \cdots, x_k is A_k^i, \cdots, x_M is A_M^i, \quad THEN y is B^i \\ R_j: & IF\ x_1 is A_1^j, \cdots, x_k is A_k^j, \cdots, x_M is A_M^j, \quad THEN y is B^j \\ O: & A_1^*, \cdots, A_l^*, \cdots, A_M^* \end{array}$$

   where A_k^i is the linguistic term of the R_i rule on the kth antecedent dimension, $k = 1, \cdots, M$. $A_l^*, l = 1, \cdots, M$ are the observed fuzzy sets of variable x_l, and M is the total number of antecedents. The distance d_k between the fuzzy sets A_k^i and A_k^* can then be calculated as:

$$d_{A_k} = \frac{d(A_k^i, A_k^*)}{sup_k - inf_k} = \frac{d(Rep(A_k^i), Rep(A_k^*))}{sup_k - inf_k} \quad (2.66)$$

   where sup_k and inf_k are the supremum and infimum of variable x_k. This normalises the absolute distance measure into the range [0,1], so that distances are compatible with others measured over different domains. From this, the distance d between a rule and an observation can then be calculated as the average of all variables' distances. The two rules which have the minimum distance are then chosen. These are located on both sides of the observation, respectively.

$$d = \sqrt{{d_{A_1}}^2 + {d_{A_2}}^2 + \cdots + {d_{A_M}}^2} \quad (2.67)$$

2. *Interpolation between the Two Rules*
Suppose that the two adjacent rules are R_i and R_j, to interpolate B^*, the values A_k^i and A_k^j are used in Eqs. 2.59 and 2.60 to obtain the displacement factor λ_{A_k}, and the intermediate fuzzy terms $A_k^{'}$ for each antecedent dimension x_k. In conjunction with the given observation terms A_k^*, the scale and move transformation $T(A_k^{'}, A_k^*)$ and the necessary parameters involved s_{A_k} and m_{A_k} are calculated using Eqs. 2.62 and 2.63.
For the current scenario with multiple antecedent rules, each antecedent dimension would have its own λ_{A_k}, s_{A_k} and m_{A_k} values. The following equations aggregate them in order to discover the intermediate fuzzy term $B^{'}$. The fuzzy set B^* of conclusion can then be estimated by the transformation $T(B^{'}, B^*) = \{s_B, m_B\}$.

$$\lambda_B = \frac{1}{M} \sum_{k=1}^{M} \lambda_{A_k} \tag{2.68}$$

$$B^{'} = (1 - \lambda_B) B^i + \lambda_B B^j \tag{2.69}$$

$$s_B = \frac{1}{M} \sum_{k=1}^{M} s_{A_k} \tag{2.70}$$

$$m_B = \frac{1}{M} \sum_{k=1}^{M} m_{A_k} \tag{2.71}$$

2.4.1.4 T-FRI with Trapezoidal Multiple Multi-antecedent Rules

1. *Determination of the Closest Rules*
Given a rule base $\mathbb{U}$, a fuzzy rule $R \in \mathbb{U}$ with M antecedents A_k, $k = 1, 2, \cdots, M$ and an observation O are expressed in the following format:

$$R: \text{IF } x_1 \text{ is } A_1, \cdots, \text{ and } x_k \text{ is } A_k, \cdots, \text{ and } x_M \text{ is } A_M, \text{ THEN } y \text{ is } B$$
$$O: A_1^*, \cdots, A_k^*, \cdots, A_M^*$$

The distance d between a rule and an observation is determined by computing the aggregated distance of all the antecedent variables:

$$d = \sqrt{\sum_{k=1}^{M} d(A_k, A_k^*)^2} \tag{2.72}$$

where

$$d(A_k, A_k^*) = \frac{d(Rep(A_k), Rep(A_k^*))}{range_k} \tag{2.73}$$

where $range_k = sup_k - inf_k$ is the domain range of the variable x_k. $d(A_k, A_k^*) \in [0, 1]$ is the normalised result of the otherwise absolute distance measure, so that distances are compatible with each other over different variable domains. The N $(N \geq 2)$ rules which have the least distance measurements with regard to the observed values A_k^*, and the conclusion B^*, are then chosen to be used in the later steps. To help explanation, assume that the observation O and a certain set of closest rules $R_i, i = 1, \cdots, N, R_i \in \mathbb{U}$ that are returned by this step are represented as follows:

$$O : A_1^*, \cdots, A_k^*, \cdots, A_M^*$$
$$R_i : \text{ IF } x_1 \text{ is } A_1^i, \cdots, \text{ and } x_k \text{ is } A_k^i, \cdots, \text{ and } x_M \text{ is } A_M^i, \text{ THEN } y \text{ is } B^i$$

2. *Construction of the Intermediate Rule*
   Let the normalised displacement factor $\omega_{A_k^i}$ denote the weight of the kth antecedent of R_i:

$$\omega_{A_k^i} = \frac{\omega'_{A_k^i}}{\sum_{i=1}^{N} \omega'_{A_k^i}} \tag{2.74}$$

   where

$$\omega'_{A_k^i} = 1/d(A_k^i, A_k^*) \tag{2.75}$$

   The intermediate fuzzy terms $A_k^\dagger$ that are to be used to build the required intermediate rule are constructed from the antecedents of the N closest rules. These are then shifted to A_k' such that they have the same representative values as those of A_k^*:

$$A_k' = A_k^\dagger + \delta_{A_k} range_{A_k}, \quad A_k^\dagger = \sum_{i=1}^{N} \omega_{A_k^i} A_k^i \tag{2.76}$$

   where the coordinates of the new fuzzy set $A_k^\dagger$ are calculated on a point-by-point basis, and δ_{A_k} is the bias between A_k^* and A_k' on the kth variable domain:

$$\delta_{A_k} = d(A_k^*, A_k^\dagger) \tag{2.77}$$

   Similar to Eq. 2.76, the shifted intermediate consequence B' can be computed, with the parameters ω_{B^i} and δ_B being aggregated from the corresponding values of A_k', such that:

$$B' = \sum_{i=1}^{N} \omega_{B^i} B^i + \delta_B range_B$$
$$\omega_{B^i} = \frac{1}{M} \sum_{k=1}^{M} \omega_{A_k^i} \tag{2.78}$$
$$\delta_B = \frac{1}{M} \sum_{k=1}^{M} \delta_{A_k}$$

3. *Scale Transformation*
For each antecedent variable of the N chosen rules, the scale transformation works by calculating two scale rates $\overline{s}_{A_k}$ and $\underline{s}_{A_k}$. The support (a_0', a_3') of the corresponding shifted fuzzy set $A^{'}$ is transformed into a new support (a_0'', a_3''), and the core (a_1', a_2') is transformed into another (a_1'', a_2''), such that:

$$\underline{s}_{A_k} = \frac{a_3'' - a_0''}{a_3' - a_0'} \tag{2.79}$$

and

$$\overline{s}_{A_k} = \frac{a_2'' - a_1''}{a_2' - a_1'} \tag{2.80}$$

This leads to a scaled fuzzy set $A_k^{''} = (a_0'', a_1'', a_2'', a_3'')$. The corresponding parameters $\underline{s}_B$ and $\overline{s}_B$ of fuzzy set B^* can be calculated as follows:

$$\underline{s}_B = \frac{1}{M} \sum_{k=1}^{M} \underline{s}_{A_k} \quad \overline{s}_B = \frac{1}{M} \sum_{k=1}^{M} \overline{s}_{A_k} \tag{2.81}$$

To maintain the convexity of a scaled fuzzy set, it is necessary to ensure that the scaled support is wider than the core. For this, the following scale ratio $\mathbb{S}$ is applied, which represents the actual increase of the ratios between the core and the support.

$$\mathbb{S} = \begin{cases} \dfrac{\frac{a_2'-a_1'}{a_3'-a_0'} - \frac{a_2-a_1}{a_3-a_0}}{1 - \frac{a_2'-a_1'}{a_3'-a_0'}} & if\ \overline{s} \geq \underline{s} \geq 0\ , \mathbb{S} \in [0, 1] \\ \dfrac{\frac{a_2'-a_1'}{a_3'-a_0'} - \frac{a_2-a_1}{a_3-a_0}}{\frac{a_2'-a_1'}{a_3'-a_0'}} & if\ \underline{s} \geq \overline{s} \geq 0\ , \mathbb{S} \in [-1, 0] \end{cases} \tag{2.82}$$

Then the $\overline{s}_B$ of consequence B^* is relevant to scale ratio $\mathbb{S}$.

$$\overline{s}_B = \begin{cases} \frac{\underline{s}_B \mathbb{S}}{\overline{s}_B} - \underline{s}_B \mathbb{S} + \underline{s}_B & if\ \overline{s}_B \geq \underline{s}_B \geq 0 \\ \underline{s}_B \mathbb{S} & if\ \underline{s}_B \geq \overline{s}_B \geq 0 \end{cases} \tag{2.83}$$

Note that for triangular fuzzy sets, the support (a_0', a_2') of the shifted fuzzy set A' is transformed into a new support (a_0'', a_2''), such that the scale rate s_{A_k} is calculated as follows:

$$s_{A_k} = \frac{a_2'' - a_0''}{a_2' - a_0'} \tag{2.84}$$

4. *Move Transformation*
   In general, for multiple antecedent rules, each variable dimension has its own move rate m_{A_k}, in order to move each of the scaled fuzzy sets A_k'' to new locations that coincide with those of the originally observed values. This allows the initially constructed intermediate fuzzy terms to completely transform. The final transformed fuzzy sets then match the exact shapes of the observed values A_k^*. Without losing generality, for a given scaled intermediate fuzzy term: $A_k'' = (a_0'', a_1'', a_2'', a_3'')$, its current support (a_0'', a_3'') and core (a_1'', a_2'') can be moved to (a_0, a_3) and (a_1, a_2), using a move rate m_{A_k} calculated as follows:

$$\begin{cases} m_{A_k} = \frac{3(a_0 - a_0'')}{a_1'' - a_0''}, & a_0 \geq a_0'' \\ m_{A_k} = \frac{3(a_0 - a_0'')}{a_3'' - a_2''}, & \text{otherwise} \end{cases} \tag{2.85}$$

   Similar to the scale transformation, the move rate m_B for the consequent dimension can be calculated by obtaining the arithmetic average of those of the antecedent variables, such that:

$$m_B = \frac{1}{M} \sum_{k=1}^{M} m_{A_k} \tag{2.86}$$

   The final interpolated result B^* can now be computed by applying the scale and move transformation to B', using the resulting parameters $\underline{s}_B$, $\overline{s}_B$ and m_B. Note that for triangular fuzzy sets, obviously, the right and centre points a_2'' and a_1'' are used (instead of a_3'' and a_2''), when computing the move ratio according to Eq. 2.85, in the case of $a_0 \leq a_0''$.

The main advantage of this method is that it always results in interpretable conclusions when the data are linguistically defined on the basis of an analogical scheme of reasoning. This method also satisfies the properties of normality and convexity. It fulfils the neighbouring structure of rules and mapping similarity as well as supporting multiple multi-antecedent rules. Its computational complexity is slightly high due to the calculations of weights, shift, scale and move parameters. In this book, T-FRI will be used as a basis for majority of the proposed BFRI approaches.

Note that extrapolation [8] is a special case of interpolation, when all of the closest rules chosen lie on one side of the given observation, the interpolation problem becomes extrapolation. Determining the closest rules and constructing the intermediate rule are carried out in the same way as those for interpolation.

One particular point which needs to be emphasised is that some attribute values of the intermediate rule probably exceed the limit of the domain space of that attribute. This is because during the construction of the intermediate rule, extrapolation may be involved and it may lead to intermediate fuzzy terms that are outside of the range. It is also possible that the fuzzified data objects exceed the domain space. Therefore, special treatment is desirable for interpolation. For general interpolation, if either the fuzzified data object or the fuzzy term of the intermediate rule exceeds the input space on a particular attribute, such an attribute is ignored when performing the interpolation as this method cannot handle this particular case.

2.4.2 Other Intermediate Rule-Based Interpolation Methods

The other techniques in this group, such as similarity transfer method [48], employ the same method for generating intermediate rules as outlined above, but the representative value is restricted to being the middle point of the core. The similarity degree is captured using the so-called lower similarity and upper similarity. By reference to the middle point of the core, a normal and convex fuzzy set can be divided into two parts, namely the lower part and the upper part. The lower similarity measures the difference of the lower parts of two fuzzy sets, by comparing the lengths of a certain *level cut*, and upper similarity does that of the upper parts. The approach in [43] ensures that the core of each fuzzy set of a created intermediate rule is equal to that of the corresponding fuzzy set for the resulting interpolated rule. In order to measure the similarity degree between two fuzzy sets with the same core, only their left slopes and the right slopes need to be compared. Two transformations, that is, increment transformation and ratio transformation, are utilised for this purpose, with one aiming to increase the length of a certain level cut of a slope during the transformation, and the other to decrease the length. A group of intermediate rule generation and firing algorithms has also been reported in [18] by means of fuzzy and semantic relations.

2.4.2.1 The ST Interpolation Method

The similarity transfer (ST) interpolation reasoning method [48] is based on the concept of a core of a fuzzy set. This method is developed to address the issue of normality and convexity that the KH method fails to guarantee. It ensures the normality and convexity for the conclusion fuzzy set if fuzzy rules involve convex and normal fuzzy (CNF) sets only. As with general fuzzy interpolation, Eqs. 2.130 and 2.134 are used to construct an intermediate rule $A \Rightarrow B$, where A has the same centre point as A^*. The ST method suggests fuzzy reasoning in the following way:

- Construct a fuzzy set A', which is near to the observation A^*. Define a certain kind of similarity between these two fuzzy sets.
- Compare A^* and A' to get their similarity, which is then transferred to the consequent part.
- Deduce B^* from B' according to the similarity transferred from the antecedent part.

The detail and the properties of the ST reasoning method can be described as follows:

Definition 2.4.1 *Similarities between two fuzzy sets:* Given two normal and convex fuzzy sets A and A' on the universes of discourse X, the lower similarity and the upper similarity between A and A' are defined as

$$S_{L(A,A')}(\alpha) = \frac{d(inf(A_\alpha), CP(A)}{d(inf(A'_\alpha), CP(A))} \tag{2.87}$$

$$S_{U(A,A')}(\alpha) = \frac{d(sup(A_\alpha), CP(A))}{d(sup(A'_\alpha), CP(A))} \tag{2.88}$$

where $\alpha \in [0, 1]$, $CP\{.\}$ represents the mid-point of core of a fuzzy set.

Then the consequence B^* is derived from the following equations:

$$CP(B^*) = CP(B') \tag{2.89}$$

$$S_{L(A^*,A')}(a) = S_{L(B^*,B')}(\alpha) \tag{2.90}$$

$$S_{U(A^*,A')}(a) = S_{U(B^*,B')}(\alpha) \tag{2.91}$$

Combining Eqs. 2.87 to 2.91 gives

$$inf(B^*_\alpha) = S_{L(A^*,A')}(\alpha) d(inf(B'_\alpha), CP(B')) + CP(B') \tag{2.92}$$

$$sup(B^*_\alpha) = S_{U(A^*,A')}(\alpha) d(sup(B'_\alpha), CP(B')) + CP(B') \tag{2.93}$$

Thus, the consequence B^* can be calculated using the representation principle of fuzzy sets.

Figure 2.10 depicts the membership functions of fuzzy sets A^1, A^2, B^1, B^2, A', B', A^* and B^*. The mid-points of their cores are a_1, a_2, b_1, b_2, a and b. The interpolation ratio β is calculated as

$$\beta = \frac{a - a_1}{a_2 - a_1} \quad \beta \in [0, 1] \tag{2.94}$$

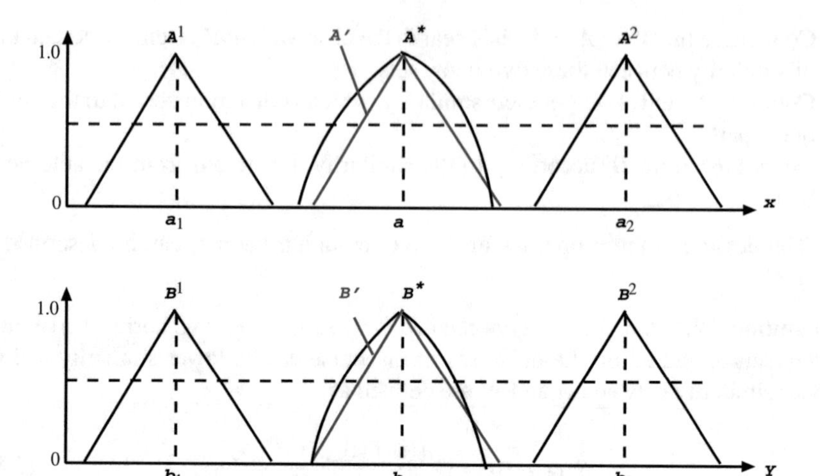

Fig. 2.10 The membership functions of fuzzy reasoning

A' is constructed as

$$inf\{A'_\alpha\} = \beta\ inf\{A^2_\alpha\} + (1-\beta) inf\{A^1_\alpha\} \tag{2.95}$$

$$sup\{A'_\alpha\} = \beta\ sup\{A^2_\alpha\} + (1-\beta) sup\{A^1_\alpha\} \tag{2.96}$$

and similarly,

$$A'_L(\alpha) = \beta\ A^2_L(\alpha) + (1-\beta)A^1_L(\alpha) \tag{2.97}$$

$$A'_R(\alpha) = \beta\ A^2_R(\alpha) + (1-\beta)A^1_R(\alpha) \tag{2.98}$$

$$B'_L(\alpha) = \beta\ B^2_L(\alpha) + (1-\beta)B^1_L(\alpha) \tag{2.99}$$

$$B'_R(\alpha) = \beta\ B^2_R(\alpha) + (1-\beta)B^1_R(\alpha) \tag{2.100}$$

also

$$CP(B') = \beta\ CP(B^2) + (1-\beta)CP(B^1) \tag{2.101}$$

that is

$$b = \beta\ b_2 + (1-\beta)b_1 \tag{2.102}$$

The similarities between A^* and A' and between B^* and B' are

$$S_{L(A^*,A')}(\alpha) = S_{L(B^*,B')}(\alpha) = \frac{A^*_L(\alpha) - a}{A'_L(\alpha) - a} \tag{2.103}$$

$$S_{U(A^*,A')}(\alpha) = S_{U(B^*,B')}(\alpha) = \frac{A^*_R(\alpha) - a}{A'_R(\alpha) - a} \tag{2.104}$$

According to the Eqs. 2.92 and 2.93, the membership function of B^* can be obtained by

$$B_L^*(\alpha) = \frac{[A_L^*(\alpha) - a][B_L'(\alpha) - b]}{A_L'(\alpha) - a} + b \tag{2.105}$$

$$B_R^*(\alpha) = \frac{[A_R^*(\alpha) - a][B_R'(\alpha) - b]}{A_R'(\alpha) - a} + b \tag{2.106}$$

From the perspective of the aforementioned properties of fuzzy rule interpolation, the ST method fulfils the neighbouring structure of rules and mapping similarity as well as supporting multiple rules and multiple antecedent variables. It has the property that the membership function of the conclusion has a shape similar to that of an observation, but also has a shape which is a linear combination of the two conclusions of the closest rules. Piecewise linearity preservation is achievable for certain cases where the membership functions are continuous.

2.4.2.2 The IRCT Method

The cutting- and transformation-based interpolation method (IRCT) [43] can be seen as a combination approach, which is based on a number of concepts, including representative value (as employed in T-FRI), collection of highest point, analogical approach using increment transformation and ratio transformation. In this method, a new rule is constructed which is closest to and has the same representative value as the observation. Then, on the basis of the similarities of the fuzzy sets in the antecedent and consequent parts, interpolative reasoning is performed with the new constructed rule, using the increment and ratio transformations.

The IRCT techniques with normal trapezoidal fuzzy sets are illustrated as below. A trapezoidal fuzzy set A can be represented by a quad, $A = (a_0, a_1, a_2, a_3)$, as shown in Fig. 2.7, where a_1 is called the "left highest point", a_2 is called the "right highest point", a_0 is called the "left bottom point" and a_3 is called the "right bottom point". The characteristic value $CV(A)$ of the trapezoidal fuzzy set is calculated as follows:

$$CV(A) = \frac{a_0 + a_1 + a_2 + a_3}{4} \tag{2.107}$$

The fuzzy interpolative reasoning scheme can be described by the following steps, shown in Fig. 2.11.

1. Using the same way to calculate λ_{rep}, let:

$$\lambda_{rep} = \frac{d(A_1, A^*)}{d(A_1, A_2)} = \frac{CV(A^*) - CV(A_1)}{CV(A_2) - CV(A_1)} \tag{2.108}$$

where $d(A_1, A^*)$ denotes the distance between the fuzzy sets A_1 and A^*, and $d(A_1, A_2)$ denotes the distance between the fuzzy sets $A1$ and $A2$.

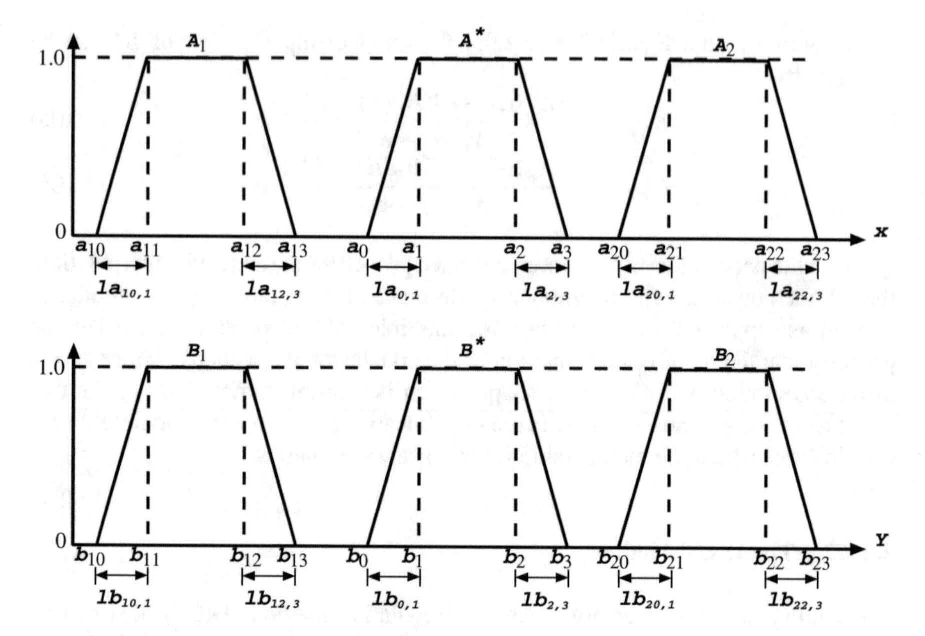

Fig. 2.11 IRCT method with trapezoidal fuzzy sets

Then, construct a new average fuzzy set A' from the fuzzy sets A_1 and A_2. Using the value of λ_{rep}, the values of $la_{0,1}'$ and $la_{2,3}'$ are calculated as follows:

$$la_{0,1}' = (1 - \lambda_{rep}) \times la_{10,1} + \lambda_{rep} \times la_{20,1} \tag{2.109}$$

$$la_{2,3}' = (1 - \lambda_{rep}) \times la_{12,3} + \lambda_{rep} \times la_{22,3} \tag{2.110}$$

Based on the values of $la_{0,1}'$ and $la_{2,3}'$, a new average fuzzy set $A' = (a_0', a_1', a_2', a_3')$ can be obtained

$$\begin{cases} a_1' = a_1 \\ a_0' = a_1' - la_{0,1}' \\ a_2' = a_2 \\ a_3' = a_2' - la_{2,3}' \end{cases} \tag{2.111}$$

2. Find the two top points b_1 and b_2 of the consequence fuzzy set B^* shown in Fig. 2.11, respectively. First, the values of λ_{a_1} and λ_{a_2} are calculated as follows:

$$\begin{aligned} \lambda_{a_1} &= \frac{d(A_1, a_1)}{d(A_1, A_2)} \\ &= \frac{a_1 - CV(A_1)}{CV(A_2) - CV(A_1)} \end{aligned} \tag{2.112}$$

$$\lambda_{a_2} = \frac{d(A_1, a_2)}{d(A_1, A_2)} = \frac{a_2 - CV(A_1)}{CV(A_2) - CV(A_1)} \tag{2.113}$$

Based on the values of λ_{a_1} and λ_{a_2}, the values of b_1 and b_2 of the consequent B^* are calculated as follows:

$$b_1 = (1 - \lambda_{a_1}) \times CV(B_1) + CV(B_2) \times \lambda_{a_1} \tag{2.114}$$

$$b_2 = (1 - \lambda_{a_2}) \times CV(B_1) + CV(B_2) \times \lambda_{a_2} \tag{2.115}$$

3. Construct a new average fuzzy set $B' = (b'_0, b'_1, b'_2, b'_3)$ from the fuzzy sets B_1 and B_2. Then, the values of $lb_{0,1}'$ and $lb_{2,3}'$ are calculated as follows:

$$lb_{0,1}' = (1 - \lambda_{rep}) \times lb_{10,1} + \lambda_{rep} \times lb_{20,1} \tag{2.116}$$

$$lb_{2,3}' = (1 - \lambda_{rep}) \times lb_{12,3} + \lambda_{rep} \times lb_{22,3} \tag{2.117}$$

a new average fuzzy set A' can be obtained

$$\begin{cases} b'_1 = b_1 \\ b'_0 = b'_1 - la_{0,1}' \\ b'_2 = b_2 \\ b'_3 = b'_2 - la_{2,3}' \end{cases} \tag{2.118}$$

4. The increment and ratio transformations are then used to compute the appropriate operators which will allow the transformation of B' to B^*.

   - *Increment transformations*: When the value of $la_{0,1}$ is greater than or equal to the value of $la_{0,1}'$, increment transformation is used. Let L_l be the increment length,

$$L_l = la_{0,1} - la_{0,1}' \tag{2.119}$$

   The value of b_0 can be calculated as

$$b_0 = b_1 - lb_{0,1} \tag{2.120}$$

   where

$$lb_{0,1} = L_l \times \frac{a_{20} - a_{13}}{b_{20} - b_{13}} + lb'_{0,1} \tag{2.121}$$

- *Ratio transformations*: When the value of $la_{0,1}$ is smaller than the value of $la_{0,1}{}'$, ratio transformation is used. Let γ_l be the ratio rate.

$$\gamma_l = \frac{la_{0,1}}{la_{0,1}{}'} \tag{2.122}$$

Then, the value of b_0 can be calculated

$$b_0 = b_1 - lb_{0,1} \tag{2.123}$$

where

$$lb_{0,1} = \gamma_l \times lb_{0,1}{}' \tag{2.124}$$

In summary,

$$b_0 = \begin{cases} b_1 - (a_1 - a_0) \times \frac{a_{20}-a_{13}}{b_{20}-b_{13}} + la_{0,1}{}' \times \frac{a_{20}-a_{13}}{b_{20}-b_{13}}, & if\ a_1 - a_0 \geq la_{0,1}{}' \\ b_1 - \frac{(a_1-a_0)\times lb_{0,1}{}'}{la_{0,1}{}'}, & if\ a_1 - a_0 < la_{0,1}{}' \end{cases} \tag{2.125}$$

In the same way, the right-angled triangle can be analysed by the same method to calculate the value of b_3 of the consequence fuzzy set B^*.

This method is quite similar to T-FRI. However, it proposes a different incremental and ratio transformation approach to obtain the conclusion using similarity measures. It is applicable to polygonal-shaped fuzzy membership functions as well as multiple multi-antecedent environments. It also satisfies the properties of normality and convexity. Its computational complexity is high due to the calculation of different parameters and transformations.

2.4.2.3 The General Methodology Method

As a member of the intermediate rule-based interpolation family, the General Methodology (GM) [18] is capable of handling arbitrary membership functions, which is its primary advantage. General interpolation claims two groups of developed algorithms: one is based on the interpolation of fuzzy relations, and the other is based on the interpolation of semantic relations. This subsection discusses the original and most typical method of this family, which consists of two key techniques: the solid cutting and the revision principle. Solid cutting [50] is used to obtain an intermediate fuzzy set A' if the observation A^* and two fuzzy rules $A_1 \Rightarrow B_1$and $A_2 \Rightarrow B_2$ are given. A ratio of λ $(0 \leqslant \lambda \leqslant 1)$ is calculated to represent the important impact of A_2 upon the construction of intermediate rule antecedent A' with respect to A_1. The solid cutting method uses the centre point of the fuzzy set to represent its overall location. λ thus can be computed as:

$$\lambda = \frac{d(CP(A_1), CP(A^*)}{d(CP(A_1), CP(A_2)} \tag{2.126}$$

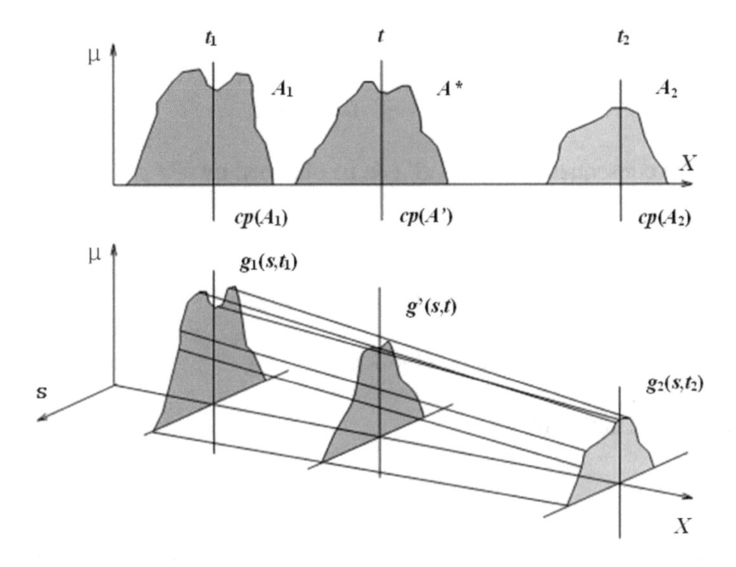

Fig. 2.12 General fuzzy interpolation

where $d(.,.)$ stands for the distance between centre points of two fuzzy sets. In the extreme cases: if $\lambda = 0$, A_2 plays no part in constructing A', while if $\lambda = 1$, A_2 plays a full role in determining A'. Figure 2.12 shows how to calculate A' if A_1, A_2 and λ are given. Dimension S is orthogonal to plane $\mu \times X$. Let $g_k(s, t_k)$, $s \in S$, $t_k = CP(A_k)$, and $k = 1, 2$, be the function that is obtained by rotating the membership function A_k: $\mu_{A_k}(x)$ by 90° around the axis t_k: $g_k(x?t_k, t_k) = \mu_{A_k}(x)$. Let a solid be constructed by fitting a surface on generatrices $g_k(s, t_k)$. Let $g'(s, t)$ be the cross section of this imagined solid at position $t = CP(A')$, where $CP(A') = \Gamma(CP(A_1), CP(A_2),)$ and $\Gamma()$ stand for the linear interpolation of two points.

The linear interpolation of two points x_1 and x_2 is:

$$x' = \Gamma(x_1, x_2, \lambda) = (1 - \lambda)x_1 + \lambda x_2, \ \lambda \in [0, 1]$$

Turning back $g'(s, t)$ into its original position, the interpolated fuzzy set A^i : $\mu_{A'}(x) = g'(x - cp(A'), cp(A'))$ is obtained.

For fuzzy interpolations only concerning triangular fuzzy sets, the solid cutting method works in the same way as the linear interpolation by using λ:

$$a_0' = (1 - \lambda)a_{10} + \lambda a_{20} \tag{2.127}$$

$$a_1' = (1 - \lambda)a_{11} + \lambda a_{21} \tag{2.128}$$

$$a_2' = (1 - \lambda)a_{12} + \lambda a_{22} \tag{2.129}$$

which are collectively abbreviated to

$$A' = (1 - \lambda)A_1 + \lambda A_2 \tag{2.130}$$

Similarly, the consequent fuzzy set B' can be obtained by

$$b'_0 = (1 - \lambda)b_{10} + \lambda b_{20} \tag{2.131}$$

$$b'_1 = (1 - \lambda)b_{11} + \lambda b_{21} \tag{2.132}$$

$$b'_2 = (1 - \lambda)b_{12} + \lambda b_{22} \tag{2.133}$$

with abbreviated notation:

$$B' = (1 - \lambda)B_1 + \lambda B_2 \tag{2.134}$$

In so doing, the newly derived rule $A \Rightarrow B$ involves the use of only normal and valid fuzzy sets. The fuzzy set A has the same centre of point as A^*. Given a fuzzy set $A \in F(X)$, $\underline{A} = \underline{support(A)}$ and $\overline{A} = \overline{support(A)}$ denote the lower and upper bounds of A in X. The revision principle-based technique [28, 51] is used to infer the fuzzy conclusion by the new rule and the observation: the revision function $y = \Lambda(x, p_1, p_2)$, where $x \in [\underline{x}, \overline{x}]$, $y \in [\underline{y}, \overline{y}]$, $p_1 = [p_{1,1} p_{1,2} \ldots p_{1,M}] \in R^M$, where $p_{1,1} = a$ and $p_{1,M} = b$, and $p_2 = [p_{2,1} p_{2,2} \ldots p_{2,M}] \in R^M$, where $p_{2,1} = c$ and $p_{2,M} = d$, subject to $p_{i,M} \leq p_{i,M+1}$, $i = 1, 2$. The revision function is a piecewise linear function where the linear pieces are defined by point-pairs $(p_{1,M}, p_{2,M})$. Figure 2.13 shows a revision function with $M = 4$.

In the triangular cases, the top point of the resulting fuzzy set B^* maintains the same position as that of B. That is, $b_1^* = b_1$. The left and right points are determined by the revision principle:

$$b_0^* = \Lambda(a_0^*, p_1, p_2) \tag{2.135}$$

$$b_2^* = \Lambda(a_2^*, p_1, p_2) \tag{2.136}$$

where

$$p_1 = [\underline{x}\ \underline{A'} CP(A') \overline{A'} \overline{x}] \tag{2.137}$$

$$p_2 = [\underline{y}\ \underline{B'} CP(B') \overline{B'} \overline{y}] \tag{2.138}$$

Thus, the consequent B^* can be calculated with the representation principle of fuzzy sets.

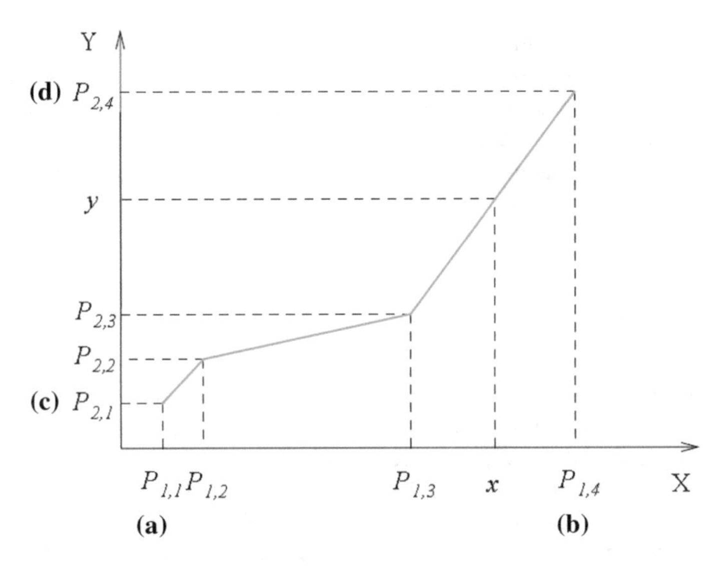

Fig. 2.13 A revision piecewise linear function

2.5 Comparison Based on General Properties

For all the outlined typical interpolation methods, comparisons are summarised based on the previous identified conditions in Table 2.1. It should be noted that the observations detailed in this table are based on observations which have been made while comparing the different techniques as part of the literature review [18, 52]. This comparison shows that most of the outlined typical approaches cannot meet all of these conditions simultaneously. In particular, the KH approach violates conditions 1 and 5 while the scale and move transformation-based approach violates conditions 3, 4 and 6. The empty entry in Table 2.1 shows that no exact empirical information

Table 2.1 Summary of comparison of typical interpolation methods based on evaluation properties

Methods	1. CNF	2. AAFMF	3. MS	4. PIB	5. CRB	6. FAR	7. AC	8. PPWL	9. MMARS
KH	F	T	T	T	F	T	T		T
HCL	T	F	T	T	T	T	T		T
FIVE		T	T	T	T	T	T		T
MACI		T	T	T	T	T	T		T
IMUL		T	T	T	T	T	T		T
T-FRI	T	T	F	F	T	F	T	T	T
GI	T		T	T	T	T	T	T	T
ST	T		T	T	T	T	T	T	T
IRCT	T	T	T		T	F	T	T	T

is available in the literature. There is still heated debate about the "properties" or "standards". Moreover, some properties may become restrictions, such as the condition 4 may limit the implementation ability of fuzzy rule extrapolation. An efficient interpolation method does not necessarily have to be fully constrained by these above characteristics. A more convincing way of evaluating them may be to apply them to real-world problems.

2.6 Summary

This chapter has presented a systematic review for the current state of the art of fuzzy rule interpolation (FRI) methods. These techniques which are able not only to perform fuzzy inferences on sparse rule bases, but also to reduce system complexity by omitting those rules which may be approximated by their neighbouring ones. Inspired by the fact that the original fuzzy interpolation approach may arrive at a result which is non-normal and convex fuzzy set, a number of modifications or improvements have been proposed in the literature. Seven of the most important approaches have been presented in this chapter.

Broadly speaking, the implementations of fuzzy interpolation can be categorised into two groups. The approaches in one group interpolate consequences directly from given observations based on a sparse rule base. The approaches in the other group first generate an intermediate rule such that the antecedent of the intermediate rule is as "close" to the given observation as possible. This intermediate rule is then fired by the given observation through similarity-based fuzzy reasoning. In particular, the scale and move transformation-based approach has been explained in detail, as this will be used as a basis for the BFRI concept proposed in this book.

On the basis of reviewing a wide range of fuzzy interpolation methods, a set of important performance evaluation criteria are identified and generalised. Although it is not necessary that *all* such criteria are fulfilled in developing and applying the existing approaches, it is expected that most of the criteria should be satisfied by a useful fuzzy rule interpolation technique, with other problem-specific parameters to fulfil given certain application. From this, several applications of fuzzy interpolation have been reviewed in the beginning of this chapter in an effort to provide an alternative but practical way of evaluation.

References

1. E.H. Mamdani, S. Assilian, An experiment in linguistic synthesis with a fuzzy logic controller. Int. J. Man Mach. Stud. 7 (1975)
2. M. Sugeno, G. Kang, Structure identification of fuzzy model. Fuzzy Sets Syst. **28**(1), 15–33 (1988)

3. T. Takagi, M. Sugeno, Fuzzy identification of systems and its applications to modeling and control. IEEE Trans. Syst. Man Cybern. **1**, 116–132 (1985)
4. T. Takagi, M. Sugeno, Comparison of fuzzy reasoning methods. Fuzzy Sets Syst. **8**(3), 253–283 (1982)
5. L. Koczy, K. Hirota, Approximate reasoning by linear rule interpolation and general approximation. Int. J. Approx. Reason. **9**(3), 197–225 (1993)
6. L. Koczy, K. Hirota, Interpolative reasoning with insufficient evidence in sparse fuzzy rule bases. Inf. Sci. **71**(1–2), 169–201 (1993)
7. Z. Huang, Q. Shen, Fuzzy interpolative reasoning via scale and move transformations. IEEE Trans. Fuzzy Syst. **14**(2), 340–359 (2006)
8. L. Koczy, K. Hirota, Fuzzy interpolation and extrapolation: a practical approach. IEEE Trans. Fuzzy Syst. **16**(1), 13–28 (2008)
9. M. Mizumoto, H.-J. Zimmermann, Comparison of fuzzy reasoning methods. Fuzzy Sets Syst. **8**(3), 253–283 (1982)
10. H. Nakanishi, I. Turksen, M. Sugeno, A review and comparison of six reasoning methods. Fuzzy Sets Syst. **57**(3), 257–294 (1993)
11. S. Kovács, Similarity based control strategy reconfiguration by fuzzy reasoning and fuzzy automata, in *Proceedings of the IEEE Annual Conference on Industrial Electronics Society*, vol. 1 (2000), pp. 542–547
12. S. Kovács, L.T. Kóczy, Application of interpolation-based fuzzy logic reasoning in behaviour-based control structures, in *Proceedings of International Conference on Fuzzy Systems*, vol. 3 (2004), pp. 1543–1548
13. S. Kovics, Fuzzy reasoning and fuzzy automata in user adaptive emotional and information retrieval systems, in *Proceedings of IEEE International Conference on Systems, Man and Cybernetics*, vol. 7 (2002), p. 6
14. K. Balázs, J. Botzheim, L. T. Kóczy, Comparative investigation of various evolutionary and memetic algorithms, in *Computational Intelligence in Engineering*. (Springer, 2010), pp. 129–140
15. Z.C. Johanyák, R. Parthiban, G. Sekaran, Fuzzy modeling for an anaerobic tapered fluidized bed reactor. Sci. Bull. Politeh. Univ. Timis. Rom. Trans. Autom. Control Comput. Sci. **52**(66), 67–72 (2007)
16. K.W. Wong, D. Tikk, T.D. Gedeon, L.T. Kóczy, Fuzzy rule interpolation for multidimensional input spaces with applications: a case study. IEEE Trans. Fuzzy Syst. **13**(6), 809–819 (2005)
17. K.W. Wong, T.D. Gedeon, Fuzzy rule interpolation for multidimensional input space with petroleum engineering application, in *Proceedings of IFSA World Congress and 20th NAFIPS International Conference*, vol. 4 (2001), pp. 2470–2475
18. P. Baranyi, L.T. Kóczy, T.D. Gedeon, A generalized concept for fuzzy rule interpolation. IEEE Trans. Fuzzy Syst. **12**(6), 820–837 (2004)
19. B. Bouchon-Meunier, R. Mesiar, C. Marsala, M. Rifqi, Compositional rule of inference as an analogical scheme. Fuzzy Sets Syst. **138**(1), 53–65 (2003)
20. Z.C. Johanyák, S. Kovács, A brief survey and comparison on various interpolation based fuzzy reasoning methods. Acta Polytech. Hung. **3**(1), 91–105 (2006)
21. L.T. Koczy, S. Kovács, Linearity and the cnf property in linear fuzzy rule interpolation, in *Proceedings of the Third IEEE Conference on Fuzzy Systems, 1994. IEEE World Congress on Computational Intelligence*. (IEEE, 1994), pp. 870–875
22. S. Yan, M. Mizumoto, W.Z. Qiao, Reasoning conditions on koczy's interpolative reasoning method in sparse fuzzy rule bases. Fuzzy Sets Syst. **75**(1), 63–71 (1995)
23. S. Chen, Y. Chang, Fuzzy rule interpolation based on the ratio of fuzziness of interval type-2 fuzzy sets. Expert Syst. Appl. **38**(10), 12 202–12 213 (2011)
24. L. Lee, S. Chen, Fuzzy interpolative reasoning using interval type-2 fuzzy sets. New Front. Appl. Artif. Intell. **5027**, 92–101 (2008)
25. D.T.I.J.L.K.P.V.B.M.T. Gedeon, Stability of interpolative fuzzy kh controllers. Fuzzy Sets Syst. **125**(1), 105–119 (2002)
26. R.C. Lee, Fuzzy logic and the resolution principle. J. ACM (JACM) **19**(1), 109–119 (1972)

27. J. Robinson, A machine-oriented logic based on the resolution principle. J. ACM (JACM) **12**(1), 23–41 (1965)
28. Z. Shen, L. Ding, M. Mukaidono, Fuzzy resolution principle, in *Proceedings of the Eighteenth International Symposium on Multiple-Valued Logic*. (IEEE, 1988), pp. 210–215
29. L. Zadeh, Fuzzy sets. Inf. Control **8**(3), 338–353 (1965)
30. L. Koczy, K. Hirota, Fuzzy logic and approximate reasoning. Synthese **30**(3–4), 407–428 (1975)
31. W. Hsiao, S. Chen, C. Lee, A new interpolative reasoning method in sparse rule-based systems. Fuzzy Sets Syst. **93**(1), 17–22 (1998)
32. P. Baranyi, D. Tikk, T.D. Gedeon, L.T. Kóczy, α-cut interpolation technique in the space of regular conclusion, in *Proceedings of IEEE International Conference on Fuzzy Systems*, vol. 1 (2000), pp. 478–482
33. P. Baranyi, D. Tikk, Y. Yam, L.T. Kóczy, L. Nadai, A new method for avoiding abnormal conclusion for α-cut based rule interpolation, in *Proceedings of IEEE International Conference on Fuzzy Systems*, vol. 1 (1999), pp. 383–388
34. D. Tikk, P. Baranyi, Comprehensive analysis of a new fuzzy rule interpolation method. IEEE Trans. Fuzzy Syst. **8**(3), 281–296 (2000)
35. D. Tikk, P. Baranyi, T.D. Gedeon, L. Muresan, Generalization of the rule interpolation method resulting always in acceptable conclusion. Tatra Mt. Math. Publ **21**, 73–91 (2001)
36. D. Tikk, P. Baranyi, L.T. Kóczy, T.D. Gedeon, On a stable and always applicable interpolation method, in *Proceedings of IEEE International Conference on Fuzzy Systems*, vol. 2 (2000), pp. 1049–1051
37. D. Tikk, P. Baranyi, Y. Yam, L.T. Kóczy, On the preservation of piecewise linearity of a modified rule interpolation approach, in *Proceedings of the EUROFUSE-SIC Conference* (1999), pp. 550–555
38. L. Koczy, K. Hirota, Stability of a new interpolation method, in *Proceedings of IEEE International Conference on Systems, Man, and Cybernetics*, vol. 3 (1999), pp. 7–9
39. Y. Yam, L. Kóczy, Representing membership functions as points in high-dimensional spaces for fuzzy interpolation and extrapolation. IEEE Trans. Fuzzy Syst. **8**(6), 761–772 (2000)
40. T.D.G.K.W. Wong, D. Tikk, An improved multidimensional α-cut based fuzzy interpolation technique, *Conf Artificial Intelligence in Science and Technology (AISAT'2000)* (2000), pp. 29–32
41. L.T.K. Sz, Kovács, Application of an approximate fuzzy logic controller in an agv steering system, path tracking and collision avoidance strategy. Fuzzy Set Theory Appl. Tatra Mt. Math. Publ., Math. Inst. Slovak Acad. Sci. **16**, 456–467 (1999)
42. T. Deng, Y. Chen, W. Xu, Q. Dai, A novel approach to fuzzy rough sets based on a fuzzy covering. Inf. Sci. **177**(11), 2308–2326 (2007)
43. S. Chen, Y. Ko, Fuzzy interpolative reasoning for sparse fuzzy rule-based systems based on α-cuts and transformations techniques. IEEE Trans. Fuzzy Syst. **16**(6), 1626–1648 (2008)
44. L. Koczy, K. Hirota, Preserving piece-wise linearity in fuzzy interpolation, in *Proceedings of IEEE International Conference on Fuzzy Systems* (2009), pp. 575–580
45. S. Jenei, Interpolation and extrapolation of fuzzy quantities revisited—(i) an axiomatic approach. Soft. Comput. **5**, 179–193 (2001)
46. S. Jenei, E.-P. Klement, R. Konzel, Interpolation and extrapolation of fuzzy quantities-the multiple-dimensional case. Soft. Comput. **6**(3–4), 258–270 (2002)
47. L. Koczy, K. Hirota, Fuzzy rule interpolation based on polar cuts, in *Computational Intelligence, Theory and Applications*. (Springer, 2006), pp. 499–511
48. M.M.S. Yan, W.Z. Qiao, An improvement to kóczy and hirota's interpolative reasoning in sparse fuzzy rule bases. Int. J. Approx. Reason. **15**, 185–201 (1996)
49. L. Ughetto, D. Dubois, H. Prade, Fuzzy interpolation by convex completion of sparse rule bases, in *Proceedings of International Conference on Fuzzy Systems* (2000), pp. 465–470
50. P. Baranyi, T.D. Gedeon, L.T. Kóczy, A general interpolation technique in fuzzy rule bases with arbitrary membership functions, in *Proceedings of IEEE International Conference on Systems, Man, and Cybernetics*, vol. 1 (1996), pp. 510–515

51. L. Ding, Z. Shen, M. Mukaidono, Revision principle for approximate reasoning, based on linear revising method, in *Proceedings of the 2nd International Conference on Fuzzy Logic and Neural Networks* (1992), pp. 305–308
52. D. Tikk, Z. Csaba Johanyák, S. Kovács, K.W. Wong, Fuzzy rule interpolation and extrapolation techniques: criteria and evaluation guidelines. J. Adv. Comput. Intell. Intell. Inform. **15**(3), 254–263 (2011)

Chapter 3
Transformation-Based Backward Fuzzy Rule Interpolation with a Single Missing Antecedent Value

Fuzzy rule interpolation significantly improves the robustness of fuzzy reasoning. It provides a way to reduce the complexity of fuzzy systems by omitting those rules which can be approximated by their neighbouring ones. Also, it can improve the applicability of fuzzy systems by allowing a certain conclusion to be generated even if the existing rule base does not cover a particular observation.

Despite the numerous proposed approaches, FRI techniques are relatively rarely applied in practice [1]. One of the main reasons for this is that many applications involve multiple-input and/or multiple-output problems. Rules are typically irregular in nature (i.e., they do not always address same antecedents). In particular, rules may be arranged in an interconnected mesh, where observations and conclusions in between different subsets of rules could be overlapped, and yet not directly related throughout the entire rule base. For such complex systems, any missing values in a given set of observations may lead to failure in interpolation. In Fig. 3.1, $R_i, i = 1, \cdots, n$ form the rule base, including interpolated rules; $x_p, x_q, p, q = 1, \cdots, m$ are the variables covering antecedents and consequence. A_q^i $(q = 1, \cdots, m, i = 1, \cdots, n)$ is the fuzzy set of the qth dimension, which is included in the ith rule. The final conclusion B^n of rule R_n cannot be interpolated in a straightforward fashion, because the three missing observations A_p^n, A_r^n and A_m^n cannot be deduced by conventional means.

For instance, consider a practical scenario in detecting terrorist bomb threats. The *Explosion likelihood* may be directly related to the *Crowdedness* of a particular location and the *Safety precautions* that are in place. The number of people in an area may be affected by the *Popularity* of the location, the level of *Travel convenience* and the amount of *Safety precautions*. A hierarchical structure for this scenario is shown in Fig. 3.2. For traditional forward interpolative reasoning, in order to interpolate *Explosion likelihood*, the observed values for *Crowdedness* and *Safety precautions* must both be provided. The variable *Safety precautions* is particularly important, as without it, no matter what other information is available, forward interpolation would still fail. Therefore, the interpolation of such crucial missing values may become necessary, in order to allow the required inference to be performed.

S. Jin et al., *Backward Fuzzy Rule Interpolation*,
https://doi.org/10.1007/978-981-13-1654-8_3

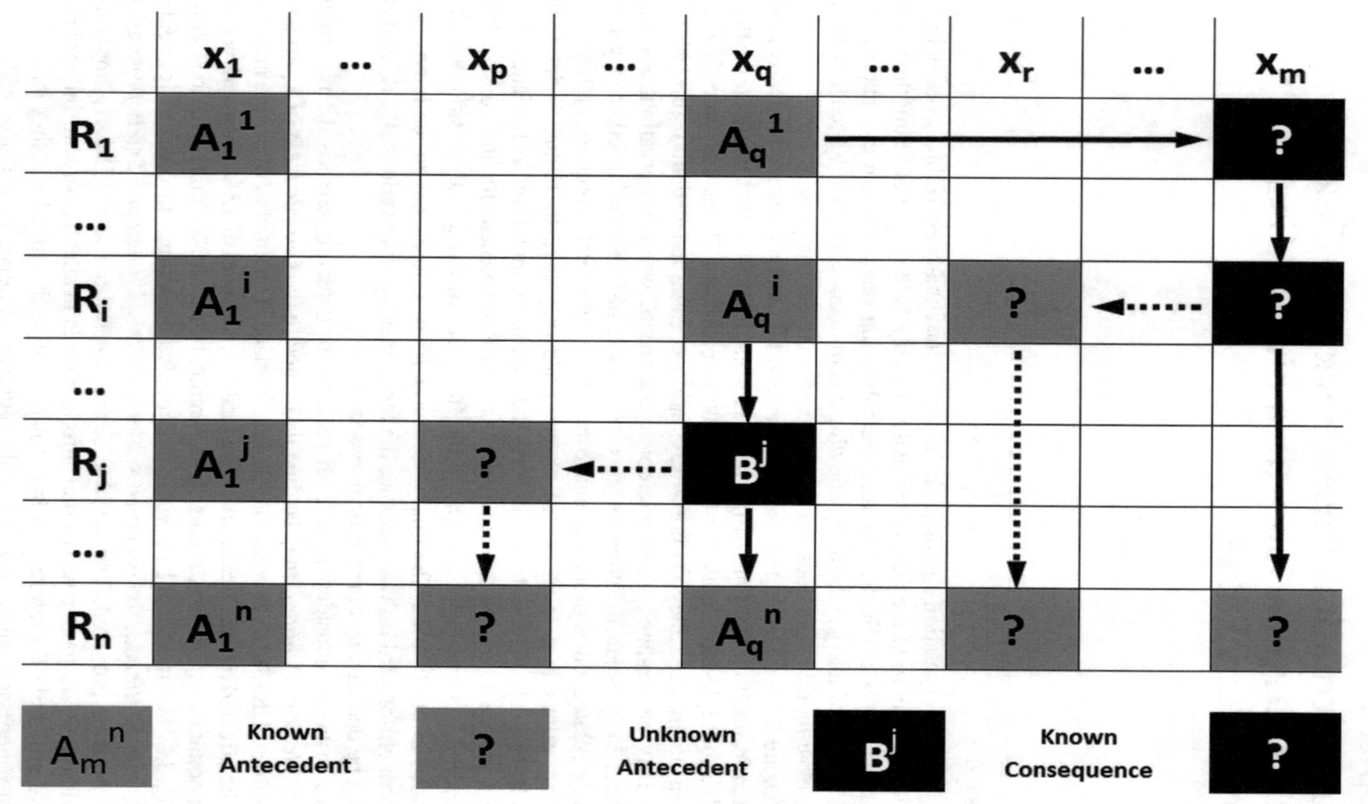

Fig. 3.1 An example system structure that may benefit from backward fuzzy rule interpolation (This hierarchical system includes multiple interrelated subsystems, with one final consequence and multiple subconsequences.)

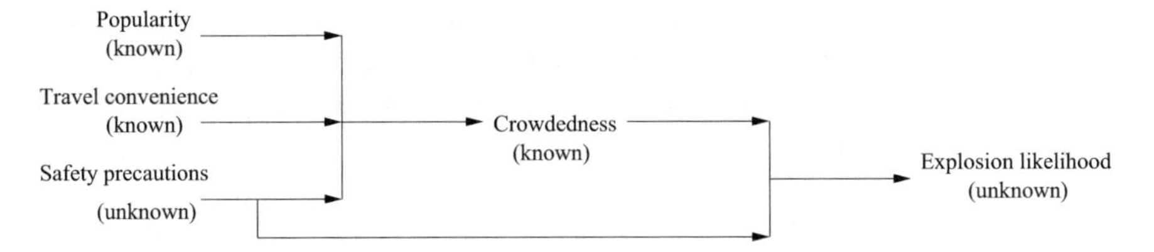

Fig. 3.2 Hierarchical fuzzy reasoning structure for terrorist bombing threat example

To address such problems, this chapter proposes a novel approach termed backward fuzzy rule interpolation and extrapolation (BFRI). This approach enables unknown antecedent values to be interpolated, given other antecedents and the conclusion. Using the example of Fig. 3.1, the unknown antecedents A_p^n and A_r^n can be backward interpolated according to rules R_j and R_i, where the conclusions B^j, B^i and the other terms are known. The last missing antecedent value A_m^n can then be interpolated using R_1, and subsequently B^n can also be computed, as now all required antecedents are known to perform forward interpolation. As such, the proposed techniques support flexible interpolation when certain antecedents are missing from the observation, where traditional FRI methods fail. In addition, BFRI also enables indirect interpolative reasoning, which involves several fuzzy rules, each with multiple antecedents. Therefore, it offers a means to broaden the application of fuzzy rule interpolation and fuzzy inference.

The remainder of this chapter is organised as follows. Sect. 3.1 introduces the general concepts of BFRI. Section 3.2 presents the BFRI approach with single missing antecedent value in detail, including both interpolation and extrapolation methods. Section 3.3 provides worked examples through the use of BFRI, demonstrating the efficacy of the proposed approach. Finally, Sect. 3.4 summarises the chapter.

3.1 General Concept of BFRI with Single Missing Antecedent Value

BFRI with single missing antecedent value (S-BFRI) is proposed for interpolation involving situations where the consequent value is known and the values of all but one antecedent variable are also given. The task is to estimate the value of that single unknown antecedent. Without loss of generality, suppose that a conventional FRIE is represented as follows:

$$B^* = f_{FRI}((A_1^*, \cdots, A_l^*, \cdots, A_M^*), (R_i), i = 1, \cdots, N) \tag{3.1}$$

where f_{FRI} denotes the interpolation/extrapolation process from M observed values, using a set of selected rules $R_i, i = 1, \cdots, N$, that are closest to $\{A_l^*|l = 1, 2, \cdots, M\}$, and B^* is the interpolated conclusion. S-BFRI can then be defined in the following form:

$$A_l^* = f_{S-BFRI}((B^*, A_1^*, \cdots, A_{l-1}^*, A_{l+1}^*, \cdots, A_M^*), (R_i), i = 1, \cdots, N) \quad (3.2)$$

where f_{S-BFRI} denotes the entire process of obtaining A_l^*, the unknown (or required) observation, which is to be backward interpolated. It uses the N closest rules, with regard to the observed (or predicted) values from the $(M - 1)$ antecedents and the conclusion B^*.

3.2 Process of S-BFRI

A close examination of the T-FRI algorithm reveals that, in order to successfully backward interpolate the missing value, a number of closest rules need to be identified first. All of the parameters involved in T-FRI (for trapezoidal fuzzy sets): ω, δ, $\underline{s}$, $\overline{s}$, $\mathbb{S}$ and m also need to be computed for the known antecedent variables, and the now observed consequent variable. The acquisition of these essential parameters allows a possible transformation process to be derived, which then helps to restore the missing antecedent value. The proposed S-BFRI algorithm that reflects this intuition is summarised below.

3.2.1 Determination of the Closest Rules

In reference to the earlier definition of S-BFRI in Eq. 3.2, when B^*, $(A_1^*, \cdots, A_{l-1}^*, A_{l+1}^*, \cdots, A_M^*)$ are given, in order to interpolate/extrapolate the unknown antecedent A_l^*, the discovery of the closest rules $R_i, i = 1, \cdots, N$, is required. Rather than using the distance measure introduced in Eq. 2.72, a modified scheme is proposed in order to reflect the biased consideration towards the consequent variable (as per the intuition indicated above):

$$d = \sqrt{w_B {d_B}^2 + \sum_{k=1,\ k \neq l}^{M} (w_{A_k} {d_{A_k}}^2)} \quad (3.3)$$

In implementing S-BFRI, without sufficient expert knowledge on the relative level of significance of different antecedents, all antecedents are treated equally:

$$w_B = \sum_{k=1}^{M} w_{A_k} = 1, \; w_{A_1} = w_{A_k} = w_{A_M} = \frac{1}{M} \tag{3.4}$$

Note that in choosing the closest rules, the square root used in the original distance measure becomes unnecessary, as only the ordering information is required. Therefore, the distance calculation can be simplified to:

$$\hat{d} = \sqrt{{d_B}^2 + \frac{1}{M-1} \sum_{k=1,\; k \neq l}^{M} {d_{A_k}}^2} \tag{3.5}$$

3.2.2 *Construction of the Intermediate Fuzzy Terms*

To aid explanation, assume a certain set of closest rules $R_i, i = 1, \cdots, N, R_i \in \mathbb{U}$ that are returned by the previous distance calculation. Following the original T-FRI algorithm, in order to create the intermediate (shifted) fuzzy terms for the known antecedent variables: A'_k, $k = 1, \cdots, M, k \neq l$, the following parameters $w_{A_k^i}, i = 1, \cdots, N$, and δ_{A_k} need to be computed first according to Eqs. 2.74–2.77. The parameter values for the intermediate (shifted) consequent fuzzy term B': w_{B^i}, $i = 1, \cdots, N$, and δ_B can be computed using exactly the same formulae as those of A_k, since its value B^* is also directly observed.

The formulae given in Eq. 2.78, although no longer needed in this scenario, reveal that both w_{B^i} and δ_B are algebraic averages of the parameter values from individual antecedent terms. For instance, if A_l was not missing, $\omega_{A_l^i}$ would become part of the sum: $\omega_{B^i} = \frac{1}{M} \sum_{k=1}^{M} \omega_{A_k^i}$ in Eq. 2.74. Thus, it has an intuitive appeal to assume that, when backward interpolating a certain parameter value for A_l, say $\omega_{A_l^i}$, the parameter value associated with the consequent variable: ω_{B^i} should be treated with a biased weight, which is the sum of all antecedent weights. The parameter values for the missing antecedent, such as $\omega_{A_l^i}$, are then calculated by subtracting those of the known antecedents from that of the consequent, as shown below:

$$\omega_{A_l^i} = M\omega_{B^i} \quad \sum_{k=1,\; k \neq l}^{M} \omega_{A_k^l} \tag{3.6}$$

Following the same logic, ω_{A_l} can be obtained:

$$\delta_{A_l} = M\delta_B - \sum_{k=1,\; k \neq l}^{M} \delta_{A_k} \tag{3.7}$$

The acquisition of these parameter values allows the construction of the intermediate (shifted) fuzzy term A_l' for the missing antecedent dimension, similar to Eqs. 2.76 and 2.78:

$$A_l' = A_l^{\dagger} + \delta_{A_l} range_{A_l} \tag{3.8}$$

where

$$A_l^{\dagger} = \sum_{i=1}^{N} \omega_{A_l^i} A_l^i \tag{3.9}$$

Note that according to the characteristics of the T-FRI algorithm, this shifted fuzzy term A_l' also determines the representative value of the final interpolated output A_l^*, since the later transformations will not alter $Rep(A_l')$.

3.2.3 Scale and Move Transformation

Having obtained the intermediate (shifted) fuzzy terms, the essential parameters $\underline{s}_{A_l}$, $\overline{s}_{A_l}$ (or a single scale rate s_{A_k} for triangular representation) and m_{A_l} involved in the transformation process can be derived. Following the same intuition and computational steps as those for $w_{A_k^i}$, $i = 1, \cdots, N$, and δ_{A_l}, by reversing the forward transformation procedure introduced in Eqs. 2.81 and 2.86, the required values can be found as follows:

$$\underline{s}_{A_l} = M\underline{s}_B - \sum_{k=1,\ k \neq l}^{M} \underline{s}_{A_k} \tag{3.10}$$

$$\overline{s}_{A_l} = M\overline{s}_B - \sum_{k=1,\ k \neq l}^{M} \overline{s}_{A_k} \tag{3.11}$$

$$m_{A_l} = Mm_B - \sum_{k=1,\ k \neq l}^{M} m_{A_k} \tag{3.12}$$

where $\underline{s}_B$, $\overline{s}_B$ and m_B are immediately obtainable by resolving Eqs. 2.79, 2.80 and 2.85. Note that, in order to guarantee the transformed fuzzy sets to be convex, $\overline{s}_{A_l}$ should be fixed in terms of the scale ratio $\mathbb{S}_{A_l}$:

$$\mathbb{S}_{A_l} = M\mathbb{S}_B - \sum_{k=1,\ k \neq l}^{M} \mathbb{S}_{A_k} \tag{3.13}$$

where $\mathbb{S}_{A_l}$ is the fixed scale ratio of A_l, $\mathbb{S}_B$ is the scale ratio of consequent dimension B, and $\mathbb{S}_{A_k}$ is the scale ratio of A_k, $k = 1, 2, \cdots, M, k \neq l$.

$$\overline{s}_{A_l} = \begin{cases} \frac{\underline{s}_{A_l} * \mathbb{S}_{A_l}}{\overline{s}_{A_l}} - \underline{s}_{A_l} * \mathbb{S}_{A_l} + \underline{s}_{A_l} & if\ \overline{s}_{A_l} \geq \underline{s}_{A_l} \geq 0 \\ \overline{s}_{A_l} * \mathbb{S}_{A_l} & if\ \underline{s}_{A_l} \geq \overline{s}_{A_l} \geq 0 \end{cases} \tag{3.14}$$

Finally with all parameters acquired, the transformation on $A_l^{'}$ can be performed, resulting in the (backward) interpolated value A_l^*.

$$T(A_l^{'}, A_l^*) = \{\underline{s}_{A_l}, \overline{s}_{A_l}, \mathbb{S}_{A_l}, m_{A_l}\} \tag{3.15}$$

3.3 Worked Examples

This section provides four worked examples of the proposed BFRI approach. For each of these, the value of the consequent variable is obtained by utilising the T-FRI method (following the forward FRI procedure of [2]), using randomly chosen values for the antecedent variables. The "missing" value is then (purposefully) removed from the observation, allowing the application of BFRI. The aim of demonstrating these examples is two-folded: (1) to show the correctness of the BFRI method; i.e., the proposed procedure can indeed restore the originally observed value; and (2) to show that the proposed distance measure is effective in identifying relevant rules in order to perform interpolation (noting that the rules involved in the initial generation process may or may not be selected).

Example 3.3.1 S-BFRI with Two Single-Antecedent and Single-Consequent Rules

An example is used here to illustrate the process of BFRI with a singleton antecedent, and also to provide an example for the *T-FRI* procedures. The two original rules are given in Table 3.1 and illustrated in Fig. 3.3, with the conclusion being given as: $B^* = (5.83, 6.26, 7.38)$.

1. *Construct Intermediate Fuzzy Terms $B^{'}$ and $A^{'}$*: The relative placement factor $\omega_B = 0.481$ is calculated first using Eq. 2.74. The intermediate fuzzy term $B^{'} = (4.811, 6.330, 8.330)$ is constructed according to Eq. 2.76, which has the same representative value as the conclusion B^*, $Rep(B^{'}) = Rep(B^*) = 6.490$. According to Eq. 2.78, work out the intermediate value $A^{'} = (5.292, 8.849, 9.849)$.
2. *Calculate Scale Rate*: The scale rate s_B is calculated using Eq. 2.62, resulting in $s_B = 0.440$. The second intermediate term $B^{''} = (5.75, 6.42, 7.30)$ denotes the fuzzy set generated by the scale transformation. This transformation rescales the support of $B^{'}$, $(b_0^{'}, b_2^{'}) = (4.81, 8.33)$ into a new support $b_0^{''}, b_2^{''} = (5.75, 7.30)$, such that the length of support is modified by s_B: $(7.30 - 5.75) = 1.55 = 0.440 \times (8.33 - 4.81)$.

Table 3.1 Example of *BFRI*, $B^* = (5.83, 6.26, 7.38)$

	O	R_1	R_2
x	A^* = missing	$A^1 = (0, 5, 6)$	$A^2 = (11, 13, 14)$
y	$B^* = (5.83, 6.26, 7.38)$	$B^1 = (0, 2, 4)$	$B^2 = (10, 11, 13)$

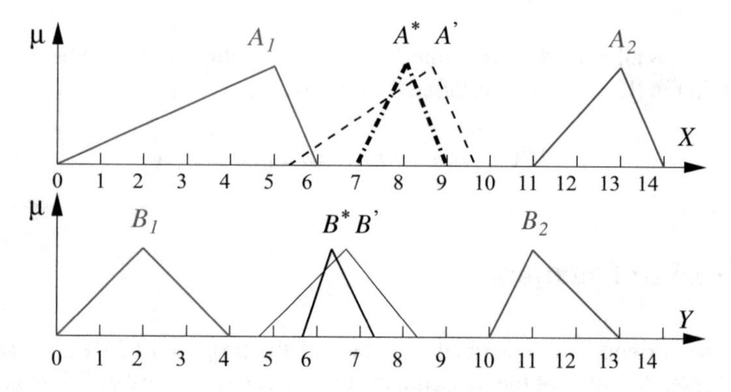

Fig. 3.3 Example of *BFRI* with single antecedent

3. *Calculate Move Ratio*: According to Eq. 2.86, $m_B = 0.357$ can be deduced that will shift $(b_0^{''}, b_2^{''}) = (5.75, 7.30)$ to $(b_0^*, b_2^*) = (5.83, 7.38)$. The result of the above scale and move transformation should successfully transform $B^{''}$ back to B^*.
4. *Scale and Move Transformation $A^{'}$ to A^**: Having discovered s_B and m_B, the reverse transformation can be performed. The scale transformation is first applied to $A^{'}$ using $s_A = s_B = 0.440$, resulting in the second intermediate term $A^{''} = (6.805, 8.372, 8.812)$. In the end, $A^* = (6.992, 7.999, 8.999)$ is a result from the move transformation from $A^{''}$ using move ratio $m_A = m_B = 0.357$, hereby completing the backward transformation process $T(A^{'}, A^*) = T(B^{'}, B^*)$.

The correctness and accuracy of backward interpolation can be easily proven by performing conventional *T-FRI* using $A^* = (6.992, 7.999, 8.999)$ as the observed value. The conclusion $B^* = (5.8304, 6.2604, 7.3803)$ is consistent with the originally given observation.

Example 3.3.2 S-BFRI with Multiple Rules and Trapezoidal Fuzzy Sets

This example illustrates S-BFRI involving multiple multiantecedent rules, where the variable values are represented by trapezoidal membership functions. The observation and the four closest rules are given in Table 3.2 and Fig. 3.4 (while the subprocess of selecting the closet rules is omitted because it is a straightforward application of Eq. 3.3 to the sparse rule base). Here, A_3^* is the missing antecedent which is to be inferred.

Table 3.2 Four closest rules for observation

	O	R_1	R_2	R_3	R_4
x_1	(3.5, 4.0, 5.0, 7.0)	(0.2, 1.1, 2.2, 2.7)	(2.0, 2.3, 2.5, 3.4)	(8.2, 9.5, 10.5, 11.0)	(10.5, 11.5, 12.5, 13.1)
x_2	(5.0, 5.5, 6.0, 7.5)	(1.5, 2.0, 2.5, 3.0)	(3.1, 3.2, 3.5, 4.3)	(7.5, 9.0, 10.2, 11.3)	(10.0, 11.2, 12.3, 13.0)
x_3	Missing	(0.4, 1.5, 2.0, 2.5)	(2.5, 3.5, 4.2, 4.5)	(7.3, 9.2, 10.5, 11.1)	(10.2, 11.0, 11.5, 13.2)
x_4	(4.5, 5.2, 6.5, 7.5)	(1.1, 1.5, 2.1, 2.5)	(6.1, 7.0, 8.0, 8.6)	(3.8, 4.1, 4.3, 5.0)	(10.1, 12.0, 12.5, 14.3)
y	(5.5, 6.5, 7.0, 8.7)	(0.2, 2.0, 2.5, 3.0)	(4.0, 4.8, 5.3, 6.0)	(9.5, 10.0, 11.3, 12.5)	(12.0, 13.0, 13.5, 14.2)

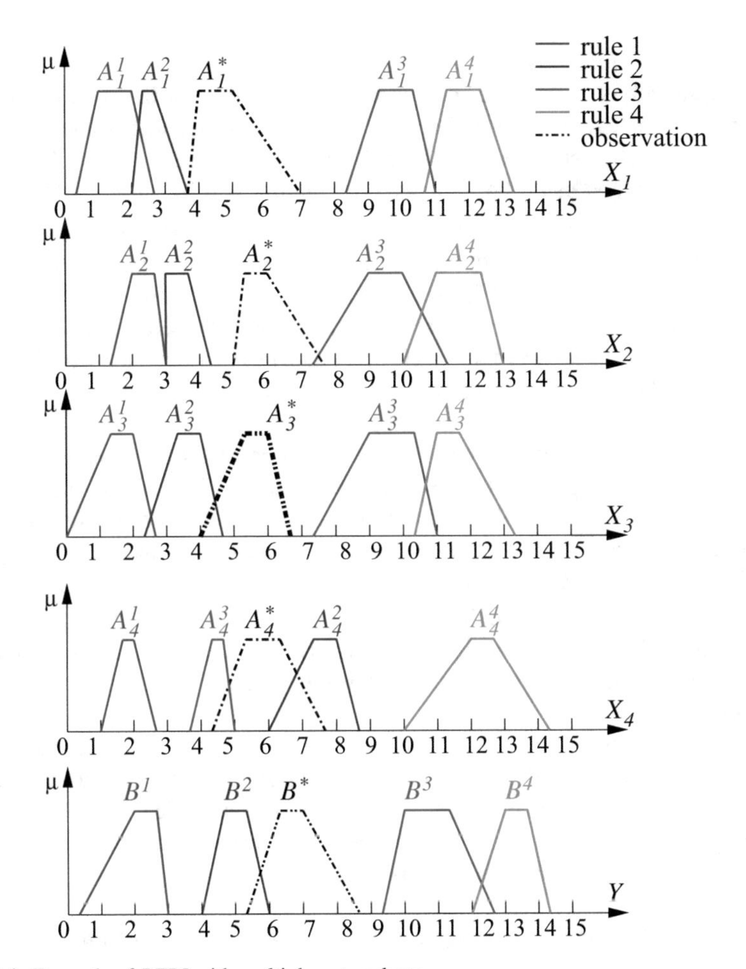

Fig. 3.4 Example of *BFRI* with multiple antecedents

Table 3.3 Normalised weights for the given antecedents

	R_1	R_2	R_3	R_4
B	0.17	0.45	0.23	0.15
A_1	0.27	0.39	0.20	0.14
A_2	0.23	0.35	0.26	0.16
A_4	0.15	0.36	0.40	0.09

1. *Construction of the Intermediate Fuzzy Terms*:
   As explained in Sect. 3.2.2, the normalised weights of the antecedents and observed conclusion are derived according to Eq. 2.74, and their values are shown in Table 3.3. The parameters for the missing observation $\omega_{A_3^i}$, $i = 1, 2, 3, 4$, can then be calculated using Eq. 3.6, resulting in: $\omega_{A_3^1} = 0.04$, $\omega_{A_3^2} = 0.70$, $\omega_{A_3^3} = 0.06$, $\omega_{A_3^4} = 0.19$. From this, the intermediate fuzzy set $A_3^\dagger = (3.39, 4.43, 5.14, 5.60)$ can be obtained according to Eq. 3.7. Then, the bias δ_{A_3} between A_3^* and $A_3^\dagger$ is calculated using Eq. 3.7, which has a value very close to 0 for this particular case, indicating that no further shifting is necessary. Therefore, the value of the shifted fuzzy term $A_3^{'} = (4.19, 5.21, 5.90, 6.49)$ can be obtained from Eq. 2.76, which has the same representative value as A_3^*.
2. *Scale and Move Transformation from $A_3^{'}$ to A_3^**:
   The individual scale and move parameters are calculated according to Eqs. 2.79–2.86, resulting in $\underline{s}_B = 1.34$, $\overline{s}_B = 0.71$, $m_B = 0.32$. The scale ratio $\mathbb{S}_{B^*} = 0.47$ is obtained using a formula similar to Eq. 3.13. Similarly, the relevant parameters $\underline{s}_{A_k}, \overline{s}_{A_k}, m_{A_k}$ of antecedents A_1^*, A_2^*, A_4^* can be obtained. Following this and using Eqs. 3.10–3.13, it can be calculated that $\underline{s}_{A_3} = 1.08$, $\overline{s}_{A_3} = 0.76$, $m_{A_3} = -0.28$ and $\mathbb{S}_{A_3} = 0.70$. The scaled fuzzy term $A_3^{''}$ is then computed to be $(4.07, 5.32, 5.84, 6.57)$. Finally, the transformed $A_3^* = (4.01, 5.46, 5.98, 6.50)$ can be obtained, which is the estimated missing value for x_3.
3. *Verification*:
   The result of BFRI can be verified by performing the conventional T-FRI, using the reconstructed observation involving A_3^*. Applying forward interpolation results in the conclusion $B^* = (5.46, 6.51, 6.85, 8.71)$, $Rep(B^*) = 6.95$. This is consistent with the given observed conclusion $(5.50, 6.50, 7.00, 8.70)$, which has a representative value of 6.98.

Example 3.3.3 S-BFRI with Triangular Fuzzy Sets and Singleton Values

To further demonstrate the generality of the proposed approach, this example illustrates S-BFRI involving multiple antecedent variables with triangular membership functions and singleton values. The two adjacent rules which involve singleton fuzzy sets are given in Table 3.4 and Fig. 3.5, with the observation being $A_1^* = (4, 5, 6)$, $A_2^* = (5, 6, 7)$, $B^* = (10, 11, 13)$.

Table 3.4 Two closest rules for observation

	O	R_1	R_2
x_1	(4, 5, 6)	(2, 2, 2)	(7, 9, 10)
x_2	(5, 6, 7)	(3, 3, 3)	(8, 9, 10)
x_3	Missing	(4, 4, 4)	(9, 10, 11)
y	(10, 11, 13)	(7, 7, 7)	(15, 17, 19)

Table 3.5 Three closest rules for observation

	O	R_1	R_2	R_3
x_1	(0, 1, 2, 3)	(3.5, 5, 6, 7)	(11, 12, 13, 14)	(7.5, 8, 9, 10)
x_2	Missing	(7, 8, 9, 10)	(4, 5, 6, 7)	(10, 11, 12, 13)
x_3	(0, 0.5, 1.5, 2.5)	(10, 11, 12, 13)	(7, 8, 9, 11)	(3, 4, 5, 6)
y	(2.0, 2.9, 3.8, 4.5)	(4.6, 5.8, 6.8, 7.9)	(8, 9, 10, 11)	(11, 12, 13, 14)

1. *Construction of the Intermediate Fuzzy Terms*:
   The parameters for the consequent dimension ω_{B^i} are calculated according to Eq. 2.74: $\omega_{B^1} = 0.57$, $\omega_{B^2} = 0.43$. The parameters for the missing observation can then be calculated using Eq. 3.6: $\omega_{A_3^1} = 0.65$, $\omega_{A_3^2} = 0.35$, and the intermediate fuzzy set $A_3^\dagger = (5.75, 6.10, 6.45)$ can be obtained with respect to Eq. 3.7. From this, using Eqs. 2.76 and 3.7, the following can be obtained: $\delta_{A_3} = 0.0$, and the shifted fuzzy set $A_3' = (5.75, 6.10, 6.45)$.
2. *Scale and Move Transformation from A_3' to A_3^**:
   The individual scale and move parameters can be calculated according to Eqs. 2.62 and 2.85, resulting in $s_B = 1.73$, $m_B = 0.33$. From Eqs. 3.10 and 3.12, it is computed that $s_{A_3} = 1.71$ and $m_{A_3} = 0.75$. The scaled fuzzy term A_3'' is therefore (5.50, 6.10, 6.70). Finally, according to Eq. 3.15, the transformed $A_3^* = (5.65, 5.80, 6.85)$ can be obtained.

Again, the result can be validated by performing the conventional T-FRI using the obtained A_3^*, resulting in the conclusion being (9.99, 11.00, 12.99), which is consistent with the given observed conclusion (10, 11, 13).

Example 3.3.4 Backward Fuzzy Rule Extrapolation

Extrapolation is a special case of interpolation, when all of the closest rules chosen lie on one side of the hyperplane in which the given observation is a certain point. Determining the closest rules and constructing the intermediate rule are carried out in the same way as those for interpolation. The example below outlines the key steps in the process of backward fuzzy rule extrapolation. Suppose that the observation and

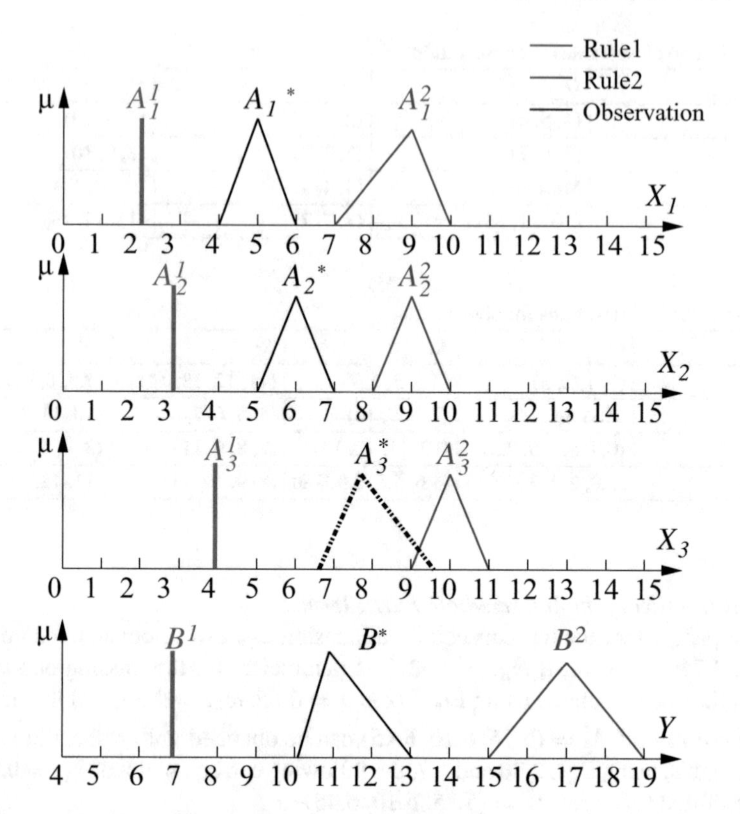

Fig. 3.5 Example of S-BFRI with triangular and singleton fuzzy sets

Table 3.6 The normalised weights for the known antecedents

	R_1	R_2	R_3
B	0.55	0.27	0.18
A_1	0.53	0.28	0.19
A_3	0.57	0.25	0.18

the three closest rules as given in Table 3.5 and Fig. 3.6 are used for extrapolation, where all rules lie on the right side of the observation. In this example, A_2^* is the missing antecedent that is to be extrapolated.

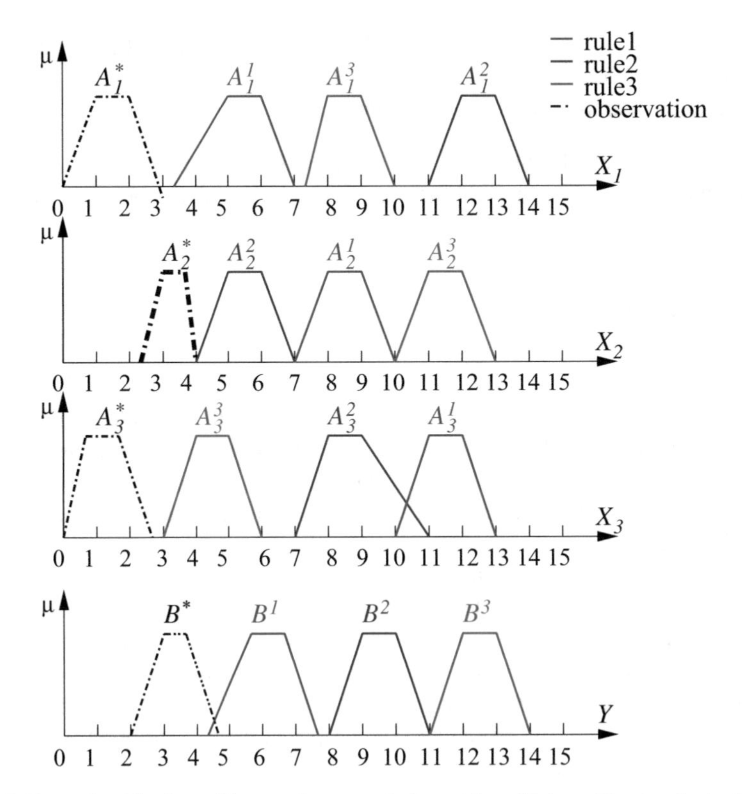

Fig. 3.6 Example of backward fuzzy rule extrapolation with multiple multiantecedent rules

1. *Construction of the Intermediate Fuzzy Terms*:
   The normalised weights associated with the observed antecedents and conclusion are listed in Table 3.6. The parameters for the missing observation $\omega_{A_2^i}$, $i = 1, 2, 3$ can then be calculated using Eq. 3.6 such that $\omega_{A_2^1} = 0.56$, $\omega_{A_2^2} = 0.27$, $\omega_{A_2^3} = 0.17$, and the intermediate fuzzy set $A_2^{\dagger} = (5.82, 6.82, 7.82, 8.82)$ can be obtained according to Eq. 3.7. Then, the bias δ_{A_2} between A_2^* and $A_2^{\dagger}$ is calculated by Eq. 3.7, $\delta_{A_2} = -0.21$. The shifted fuzzy term $A_2^{'}$, which has the same representative value as A_2^*, can be obtained from Eq. 2.76: $A_2^{'} = (2.73, 3.73, 4.73, 5.73)$.
2. *Scale and Move Transformation from $A_3^{'}$ to A_3^**:
   The individual scale and move parameters are calculated with respect to Eqs. 2.79–2.86, resulting in $\underline{s}_B = 0.64$, $\overline{s}_B = 0.59$, $m_B = -0.61$. The scale ratio $\mathbb{S}_{B^*} = 0.09$ is obtained according to Eq. 3.13. Similarly, the relevant parameters $\underline{s}_{A_k}, \overline{s}_{A_k}, m_{A_k}$ of antecedents A_1^*, A_3^* can be obtained. In particular, using equations similar to Eqs. 3.10–3.13, it follows that $\underline{s}_{A_2} = 0.40$, $\overline{s}_{A_2} = 0.50$, $m_{A_2} =$

-0.91 and $\mathbb{S}_{A_2} = -0.04$. The scaled fuzzy term A_2'' can then be computed: $(3.27, 3.94, 4.52, 5.19)$. Finally, the required $A_2^* = (3.14, 4.21, 4.79, 5.05)$ is obtained by performing the transformation.

Note that as with the previous illustrative cases for interpolation, the above-extrapolated result can be verified that it matches well with the observation.

3.4 Summary

In this chapter, a new approach for fuzzy interpolation termed backward interpolation and extrapolation (BFRI) has been presented in an attempt to address the potential weaknesses of the original FRI methods. In practical applications with interconnected subsets of rules, situations may arise when a crucial antecedent of observation is absent, either due to human error or difficulty in obtaining data, while the associated conclusion may be derived according to alternative rules or even observed directly. If such missing antecedents were involved in the subsequent interpolation process, the final conclusion would not be deduced using conventional means. However, missing antecedents may be related to certain conclusion and therefore may be inferred or interpolated using the known antecedents and conclusion. For this purpose, this chapter has presented a novel extension to traditional fuzzy rule interpolation by introducing methods to perform BFRI. The proposed method supports flexible interpolation when a certain antecedent is missing from the observation, where traditional approach fails.

The proposed technique builds upon the scale and move transformation-based fuzzy interpolation mechanism. BFRI with single missing antecedent value (S-BFRI) is for interpolation involving situations where the consequent value is known and the values of all but one antecedent variable are also given. S-BFRIcan support intertwined multiple multiantecedent rules involving triangular and trapezoidal membership functions. A modified biased distance measure scheme is proposed also. In order to successfully backward interpolate the missing value, all of the parameters involved in T-FRI are computed for the known antecedent variables, and the observed consequent variable. The acquisition of these essential parameters allows a possible transformation process to be derived, which then helps to restore the missing antecedent value. In particular, the approach supports both interpolation and extrapolation which involve multiple intertwined fuzzy rules, each with multiple antecedents. Four worked examples have been provided to illustrate the process involved in this approach. The results demonstrate that the proposed distance measure is effective in identifying relevant rules, and the proposed procedure can indeed restore the originally observed value with an desired correctness.

References

1. S. Kovács, Special issue on fuzzy rule interpolation. J. Adv. Comput. Intell. Intell. Inform. **253** (2011)
2. S. Kovács, Fuzzy interpolation and extrapolation: a practical approach. IEEE Trans. Fuzzy Syst. **16**(1), 13–28 (2008)

Chapter 4
Transformation Based Backward Fuzzy Rule Interpolation with Multiple Missing Antecedent Values

S-BFRI [1] enables a missing antecedent value to be interpolated in a backward fashion by exploiting the other given antecedents and the consequent. S-BFRI works by performing indirect interpolative reasoning which involves several intertwined fuzzy rules, each with multiple antecedents. However, no existing technique, (including BFRI) considers the case where multiple antecedents are absent. This unfortunately may be more common in practical problems such as medical diagnosis [2], network intrusion detection [3] and oil exploration [4]. Therefore, the question about how to perform BFRI with multiple missing values is raised. This chapter presents two approaches in order to deal with situations where multiple antecedent values are missing from a given observation: (1) the parametric approach (Sect. 4.1), which directly extends the S-BFRI method but involves a higher computational complexity; and (2) the feedback approach (Sect. 4.2), which is a more generalised method that works more closely with conventional FRI procedures.

The remainder of this chapter is organised as follows. Section 4.1 presents the proposed parametric approach, along with worked examples. Section 4.2 presents the proposed feedback approach, along with worked examples. Section 4.3 compares the efficacy of the these methods using simulated examples. Finally, Sect. 4.4 concludes the chapter.

4.1 Parametric Approach

The key to solving general BFRI problems, following the principles of the S-BFRI method, lies with the calculation of the best T-FRI parameter combination, which leads to the closest resemblance of the original (missing) values. In particular, to create the intermediate fuzzy terms based on the N closest rules, the following set of parameters:

$$\{(\omega_{A_l^i}, i = 1, 2 \ldots, N), \delta_{A_l}, \underline{s}_{A_l}, \overline{s}_{A_l}, m_{A_l}\} \tag{4.1}$$

of cardinality $N + 4$ is required to backward interpolate each missing antecedent A_l^*, given trapezoidal representation. Here, $l \in L$, $L \subseteq \{1, \ldots, M\}$, denotes the indices

S. Jin et al., *Backward Fuzzy Rule Interpolation*,
https://doi.org/10.1007/978-981-13-1654-8_4

of the missing antecedents A_l^*. Taking parameter for bias: δ_{A_l} as an example, the following constraint needs to be satisfied:

$$\sum_{l \in L} \delta_{A_l} = M\delta_B - \sum_{k=1,\ k \notin L}^{M} \delta_{A_k} \tag{4.2}$$

which is an extended form of Eq. 3.7 used in S-BFRI. Similar formulae for the remaining parameters may also be derived in exactly the same fashion, altogether forming multiple simultaneous equations which must then be resolved.

Note that apart from those determined by the aforementioned equations, by definition, these parameters also take values from their underlying ranges: $\omega \in [0, 1]$, $\delta \in [-1, 1]$, $\underline{s}_{A_l} \in [0, \infty)$, $\overline{s}_{A_l} \in [0, \infty)$, and $m \in [-1, 1]$. Thus, these ranges need to be discretised in order to generate the required parameter combinations. In assessing the performance of the estimation, the conventional T-FRI procedure can then be invoked to verify the correctness and accuracy. This is done by comparing the output (using the estimated parameters) with the actual observed consequent value, so that the most suitable setting may be identified.

After all parameters are acquired, in theory, the missing antecedents may be expected to be individually derived using the previously described S-BFRI steps. Unfortunately, the set of simultaneous equations cannot be resolved in a straightforward manner, due to the lack of sufficient given values. This is because different observations may potentially lead to the same (or a very similar) consequent, for any system of a fair complexity. As the number of missing antecedent values increases, the possible scenarios may become extremely wide-reaching or even countless.

From a theoretical point of view, the complexity of this approach mainly comes from the high number of possible parameter combinations (ω, δ, $\underline{s}$, $\overline{s}$, and m), all non-independent. The weight: ω, in particular, is calculated with regards to all $|L|$ missing antecedents and all N closest rules, thus resulting in a considerably high complexity: $O(\upsilon^{N|L|})$. Here $\upsilon \in \mathbb{N}^+$, $\upsilon > 1$, signifies the number of discretised intervals that are used to generate the possible parameter combinations. Higher values for υ produce finer intervals, and allows closer estimations to the actual values. The discretisations of the other three parameters: δ, $\underline{s}$, $\overline{s}$, and m all have the same computational complexity of $O(\upsilon^{|L|})$. Therefore, the overall computational complexity of generating these combinations is $O(\upsilon^{(N+4)|L|})$, which is prohibitive for large υ, N, and $|L|$. The situation worsens if the verification of every possible parameter combination is required in order to validate that the estimated transformations do indeed produce reasonable results, and to enable the selection of better (closer) outputs. The resultant P-BFRI process will have an overall cost of $O(\upsilon^{(N+4)|L|}) \cdot O(\text{FRI})$, where $O(\text{FRI})$ stands for the complexity of the FRI process itself.

Having undertaken the above analysis of the computational complexity that theoretical T-FRI involves, practically simpler methods are necessary. For this, a simplified process that supports the interpolation of two missing antecedent values is

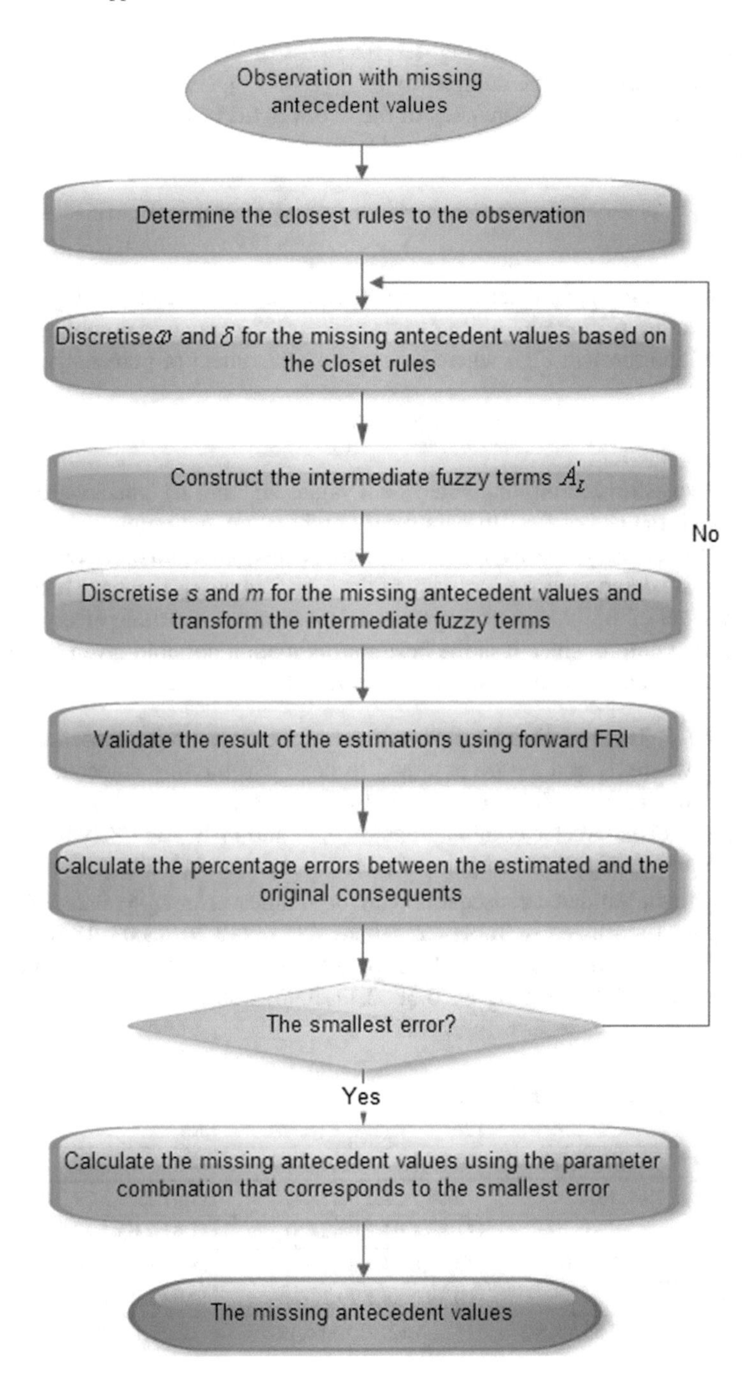

Fig. 4.1 Flowchart of the parametric approach

proposed here, and outlined in Fig. 4.1. The cost of discretising ω is reduced to $O(\upsilon^N)$, with an overall time complexity of $O(\upsilon^{N+4}) \cdot O(\text{FRI})$. This simplification takes advantage of the codependency of the two weights ($\omega_{A^i_{l_1}}$ and $\omega_{A^i_{l_2}}$), associated with each rule R^i:

$$\omega_{A^i_{l_1}} + \omega_{A^i_{l_2}} = M\omega_{B^i} - \sum_{k=1,\ k\notin L}^{M} \omega_{A^i_k} \tag{4.3}$$

where one value uniquely determines the other during the discretisation process.

The outcomes of the now more manageable υ^{N+4} parameter combinations can then be verified through FRI, where a simple measurement of percentage error $\epsilon^j_{\%}$ may be used:

$$\epsilon^j_{\%} = 100 \times d(B^{j*}, B^*) \tag{4.4}$$

Here, the two estimated missing antecedent values $A^{j*}_{l_1}$ and $A^{j*}_{l_2}$, backward interpolated through the use of the jth parameter combination, are employed to obtain a certain B^{j*}. The distance between this estimated consequent and the actual observed consequent B^* determines the accuracy of the transformations that have taken place. Finally, the set of A^{j*}_l which corresponds to the minimal resulting error is chosen as the desired output, since it is the best approximation possible given the limited information, and the amount of discretised intervals employed.

Example 4.1.1 Table 4.1 lists the observation and the closest rules chosen according to Eq. 3.5, where x_3 and x_5 are assumed to have two missing antecedent values. For this particular example, if $\upsilon = 12$ is used for discretising each parameter, the total number of parameter combinations ω, δ, s, and m is $12^{(2+4)}$ (with $N = 2$), resulting in the same number of pairs of possible A^{j*}_3 and A^{j*}_5. The parameters corresponding to the validated consequent with the smallest error $\epsilon^j_{\%} = 0.22\%$ (according to Eq. 4.4) are listed in Table 4.2, where $B^{j*} = (12.70, 13.03, 13.63, 14.66)$. The final backward interpolative values for the two missing antecedents are $A^*_3 = (2.02, 2.18, 2.40, 2.77)$, and $A^*_5 = (0.52, 2.17, 3.03, 3.55)$.

Table 4.1 Two closest rules for observation

	O	R_1	R_2
x_1	(2.88, 3.50,3.90, 4.66)	(3.07,3.55,3.73,4.08)	(3.51,3.75,3.99,4.23)
x_2	(0.96, 1.29,1.56, 1.97)	(0.88,1.21,1.32,1.47)	(1.57,2.19,2.29,3.01)
x_3	missing	(1.23,1.64,2.08,2.65)	(1.41,1.83,2.22,2.68)
x_4	(8.37,8.56, 9.24, 9.68)	(7.91,8.34,8.87,9.79)	(8.03,8.62,8.90,9.14)
x_5	missing	(0.39,0.88,1.29,1.97)	(0.94,1.41,1.92,2.34)
y	(9.56,10.54,10.83,11.58)	(11.36,12.40,12.66,12.70)	(12.77,13.83,14.54,15.19)

Table 4.2 Parameters for estimating the missing antecedents

A_l	$\omega_{A_l^1}$	$\omega_{A_l^2}$	δ_{A_l}	$\underline{s}_{A_l}$	$\overline{s}_{A_l}$	m_{A_l}
$l = 3$	0.92	0.08	0.82	0.61	0.90	−1.0
$l = 5$	0.39	0.61	−0.27	1.13	0.15	0.23

4.2 Feedback Approach

This section describes an alternative and more intuitive approach to BFRI, termed the feedback approach (F-BFRI). It significantly reduces the time-complexity for parameter estimation. This is shown in the flowchart of Fig. 4.2, and illustrated in Algorithm 4.2.1. It works by directly estimating the possible initial values of the missing antecedents, and then validating the resultant consequent through conventional FRI, in order to identify the most suitable value combination(s) that lead to the observed consequent value. For consistency and ease of explanation, as with P-BFRI, mechanisms such as T-FRI, discretisation of variable ranges, and the percentage error-based validation are again used in the implementation.

Algorithm 4.2.1: Approximation Procedure

for $l \in L$ **do**
- $o = \emptyset$
- **for** $c_l^j = \inf\{x_l\}$ **to** $\sup\{x_l\}$ **do**
  - $A_l^{j*} = approx(c_l^j)$
  - $o = o \cup A_l^i$
  - $c_l^j = c_l^j + range_{x_l}/\upsilon$
- $combinations = combinations \cup o$

return $combinations$

In order to obtain the initial estimation, the domain ranges of the missing antecedents themselves (rather than those of the parameters previously used for P-BFRI) are discretised into υ intervals. The resulting crisp points are then used to approximate a total of $\upsilon^{|L|}$ possible value combinations for the missing antecedent variables $\{A_l^{j*}\}, l \in L, j = 1, \cdots, \upsilon^{|L|}$. Assume a given crisp point c_l^j for the lth antecedent variable, Eqs. 4.5–4.7 detail this approximation procedure, which is denoted by $approx(c_l^j)$ in Algorithm 4.2.1.

$$\begin{cases} a_{0_{A_l^{j*}}} = c_l^j - \frac{\sum_{i=1}^{N}(Rep(A_l^i) - a_{0_{A_l^i}})}{N} \cdot \underline{\Delta} \\ a_{1_{A_l^{j*}}} = c_l^j - \frac{\sum_{i=1}^{N}(Rep(A_l^i) - a_{1_{A_l^i}})}{N} \cdot \overline{\Delta} \\ a_{2_{A_l^{j*}}} = c_l^j + \frac{\sum_{i=1}^{N}(a_{2_{A_l^i}} - Rep(A_l^i))}{N} \cdot \overline{\Delta} \\ a_{3_{A_l^{j*}}} = c_l^j + \frac{\sum_{i=1}^{N}(a_{3_{A_l^i}} - Rep(A_l^i))}{N} \cdot \underline{\Delta} \end{cases} \tag{4.5}$$

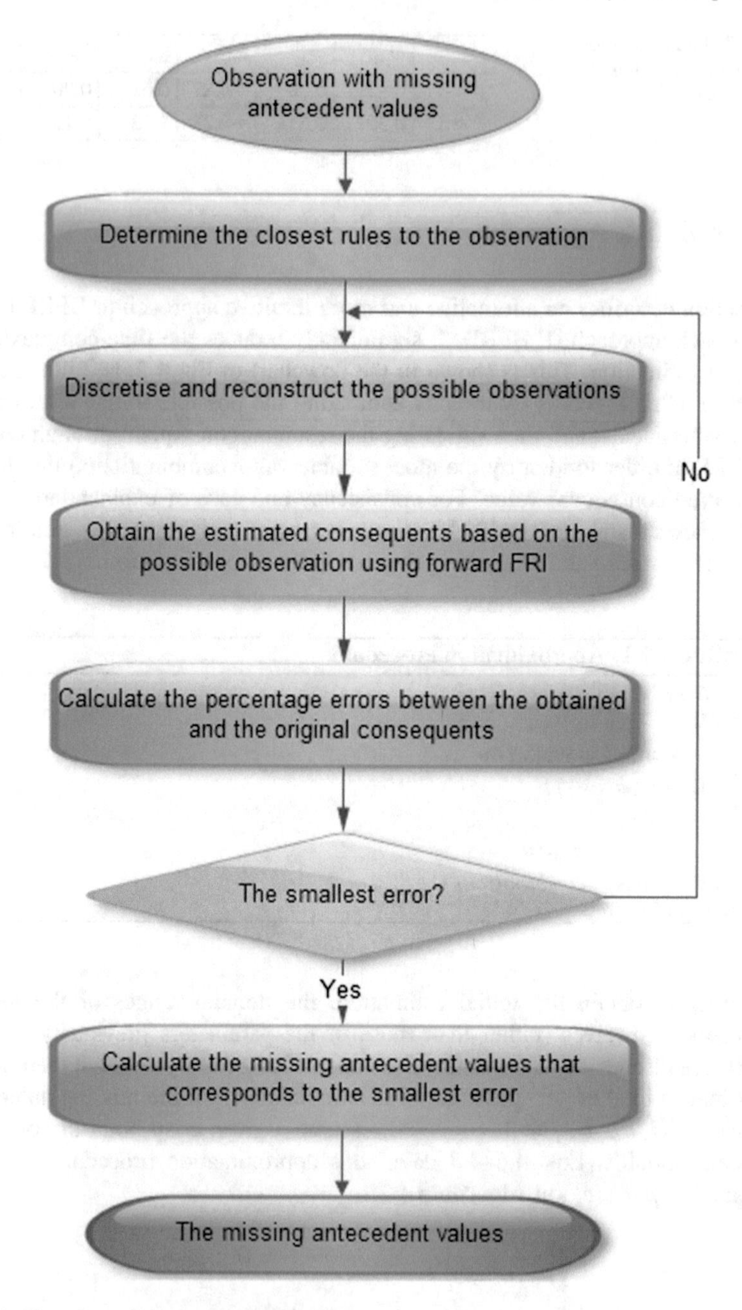

Fig. 4.2 Flowchart of the feedback approach

where

$$\underline{\Delta} = \frac{N}{M - |L| + 1} \left(\sum_{k=1,\ k \notin L}^{M} \frac{a_{3A_k^*} - a_{0A_k^*}}{\sum_{i=1}^{N} (a_{3A_k^i} - a_{0A_k^i})} + \frac{a_{3B^*} - a_{0B^*}}{\sum_{i=1}^{N} (a_{3B^i} - a_{0B^i})} \right) \quad (4.6)$$

$$\overline{\Delta} = \frac{N}{M - |L| + 1} \left(\sum_{k=1,\ k \notin L}^{M} \frac{a_{2A_k^*} - a_{1A_k^*}}{\sum_{i=1}^{N} (a_{2A_k^i} - a_{1A_k^i})} + \frac{a_{2B^*} - a_{1B^*}}{\sum_{i=1}^{N} (a_{2B^i} - a_{1B^i})} \right) \quad (4.7)$$

M is the total number of antecedent variables, and N is the number of the closest rules.

To obtain close approximations of the missing values, it is useful to remind that the estimated fuzzy sets A_l^{j*} are influenced by both the selected rules: $R_i, i = 1, \cdots, N$, and the observed values $A_k^*, k \notin L$ and B^*. In particular, for a trapezoidal fuzzy set A_l^{j*} that may be returned as the estimated outcome, the positions of its four points are defined relative to the averaged (over the N closest rules) displacements between their corresponding points and the representative values of the fuzzy antecedents $A_l^i, i = 1, \cdots, N$. The points $a_{0A_l^{j*}}$ and $a_{3A_l^{j*}}$ are then scaled with respect to the ratio $\underline{\Delta}$, calculated from the averaged (over the N closest rules and all known antecedent/consequent dimensions) ratios between the supports of the observed values, and those of the existing rules. Similarly, $a_{1A_l^{j*}}$ and $a_{3A_l^{j*}}$ are adjusted with respect to $\overline{\Delta}$.

The $\upsilon^{|L|}$ possible combinations of the fuzzy sets being estimated are used to obtain their respective consequent values $B^{j*}, j = 1, \cdots, \upsilon^{|L|}$, through the conventional T-FRI procedure. Note that the closest rules chosen for each of the combinations may be different, since the distance calculation is purely based on the values of the currently estimated observations. The percentage error $\epsilon_{\%}^{j}$ is then calculated using Eq. 4.4, and the estimated missing antecedent values corresponding to the smallest $\epsilon_{\%}^{j}$ are returned as the final result.

The computational complexity of this approach is principally due to the generation process of the initial fuzzy sets: $O(\upsilon^{|L|})$, which is much more scalable than that of P-BFRI: $O(\upsilon^{(N+4)|L|})$. The run-time cost of F-BFRI is also independent of the number of closest rules N, and of course, the overhead incurred due to the estimation of the T-FRI parameters as required in P-BFRI is also eliminated.

Example 4.2.1 Consider the observation given in Table 4.3, where the values of x_1, x_3, x_5, and x_7 are assumed to be missing. According to Eq. 4.5, a total of 20^4 ($\upsilon = 20$, $|L| = 4$) possible value combinations are used to generate the same number of potential consequent values. O^{p*} and O^{q*}, which are two of such combinations obtained in the process and the different closest rules chosen using Eq. 3.5 are shown in Table 4.3. After forward interpolation with these approximated fuzzy sets, the $\epsilon_{\%}^{j}$ obtained during this example evaluation is given in Table 4.4. The smallest $\epsilon_{\%}$ is 0.74%, and the corresponding consequent $B^{q*} = (9.74, 10.16, 10.93, 11.08)$. The resultant estimated outcome of the four missing antecedents: $A_1^* = (1.73, 2.70, 3.57, 3.97)$, $A_3^* = (2.33, 3.53, 4.51, 5.58)$, $A_5^* = (4.73, 5.71, 6.61, 7.44)$, and $A_7^* = (3.87, 4.65, 5.23, 5.79)$ is therefore the final BFRI result.

Table 4.3 Closest rules for two example reconstructed observations O^{p*} and O^{q*}, $p, q \in 1, \cdots, v^{|L|}$

	O^{p*}	R_1	R_2	R_3
x_1	$A_1^{p*} = (6.12, 6.70, 7.28, 7.55)$	(7.68,8.18,9.18,9.68)	(8.06,8.56,9.56,10.06)	(7.84,8.34,9.34,9.84)
x_2	(1.32,1.74,2.20,2.78)	(6.09,6.59,7.59,8.09)	(0.64,1.14,2.14,2.64)	(8.68,9.18,10.18,10.68)
x_3	$A_3^{p*} = (7.37, 7.95, 8.53, 9.03)$	(7.91,8.41,9.41,9.91)	(6.47,6.97,7.97,8.47)	(0.39,0.89,1.89,2.39)
x_4	(4.36,4.58,4.63,5.32)	(4.75,4.89,5.12,5.73)	(5.11,5.14,5.24,5.89)	(3.51,4.38,4.58,5.00)
x_5	$A_5^{p*} = (7.72, 8.29, 8.53, 8.70)$	(5.86,6.36,7.36,7.86)	(0.15,0.65,1.65,2.15)	(6.72,7.22,8.22,8.72)
x_6	(4.19,4.47,5.35,5.80)	(7.76,8.26,9.26,9.76)	(8.08,8.58,9.58,10.08)	(0.60,1.10,2.10,2.60)
x_7	$A_7^{p*} = (3.64, 4.22, 4.80, 5.37)$	(5.92,6.42,7.42,7.92)	(0.67,1.17,2.17,2.67)	(1.52,2.02,3.02,3.52)
y		(5.79,6.29,7.29,7.79)	(7.73,8.23,9.23,9.73)	(0.31,0.81,1.81,2.31)
	O^{q*}	R_1	R_2	R_3
x_1	$A_1^{q*} = (1.73, 2.70, 3.57, 3.97)$	(1.00,1.55,2.27,2.78)	(5.03,5.73,6.26,7.01)	(8.29,8.33,8.70,9.09)
x_2	(1.32,1.74,2.20,2.78)	(1.44,1.87,2.38,3.00)	(3.74,4.24,4.56,5.02)	(3.50,4.37,4.83,5.25)
x_3	$A_3^{q*} = (2.33, 3.53, 4.51, 5.58)$	(8.72,9.06,9.65,10.04)	(7.57,7.99,8.46,8.84)	(7.44,7.98,8.58,8.90)
x_4	(4.36,4.58,4.63,5.32)	(4.75,4.89,5.12,5.73)	(5.11,5.14,5.24,5.89)	(3.51,4.38,4.58,5.00)
x_5	$A_5^{q*} = (4.73, 5.71, 6.61, 7.44)$	(6.47,6.89,7.53,8.01)	(4.17,4.94,5.67,6.32)	(5.57,5.95,6.38,7.17)
x_6	(4.19,4.47,5.35,5.80)	(6.47,6.89,7.53,8.08)	(5.77,5.86,6.34,7.24)	(5.32,5.84,6.74,7.77)
x_7	$A_7^{q*} = (3.87, 4.65, 5.23, 5.79)$	(1.78,2.68,3.58,4.12)	(4.91,5.51,6.31,6.87)	(6.01,6.07,6.26,7.33)
y		(8.00,9.04,9.64,10.08)	(6.09,6.67,7.19,7.34)	(4.79,5.57,6.31,6.62)

Table 4.4 Errors $\epsilon_{\%}$ between B^{p*}, B^{q*} and B^*

Consequent	Value	$\epsilon_{\%}$
B^*	(8.46,9.26,9.53,9.83)	
⋮	⋮	⋮
B^{p*}	(9.74,10.16,10.93,11.08)	6.14%
⋮	⋮	⋮
B^{q*}	(8.09,8.54,9.60,10.10)	0.74%
⋮	⋮	⋮

4.3 Comparative Studies

To systematically compare the two proposed methods: P-BFRI and F-BFRI, a numerical function shown in Eq. 4.8 is used. The rule base employed in the experimental evaluation is generated using the following steps: (1) a random set of crisp values are selected for the function variables, and the outcome is calculated according to Eq. 4.8; (2) these crisp values are then fuzzified into trapezoidal fuzzy sets; and (3) the rule base is then populated using these randomly generated rules, while checking (and where appropriate, removing) rules to ensure that the underlying domain is reasonably covered, whilst there still exist sufficient gaps between rules in order to utilise interpolation.

This experimental set up enables an initial sparse rule base to be generated that is an approximation of the underlying knowledge, simulating those obtainable by "subject experts". An observation is obtained in a similar manner, where the "missing" values are then purposefully removed to facilitate backward reasoning. Since the underlying function, i.e., "ground truth", is available. The consistency, accuracy, and robustness of the interpolative procedure can therefore be verified. This is done by comparing the outcome of the interpolation to the actual value computed using Eq. 4.8.

$$y = 3x_1 - 3.3x_2 + 0.4x_3 + 0.5x_4 + 0.7x_5 \tag{4.8}$$

This test therefore, reflects an underlying principle similar to that behind cross validation and statistical evaluation [5, 6].

Altogether, 500 simulated samples are randomly drawn from the domain $U = [0, 10]^5$. Without loss of generality, the values of x_3 and x_5 are assumed to be missing. The errors for the consequent and the missing antecedents over the 500 testing records are illustrated in Figs. 4.3 and 4.4 and summarised in Table 4.5. For the consequent variable B, the errors are obtained by calculating the distance between the estimated consequent B^{j*} and the actual value B^*. The errors of the two antecedent variables with missing values: A_3 and A_5 are derived from the distances between the interpolative outcomes (i.e., values corresponding to the smallest consequent error) and the actual values of A_3^* and A_5^* (the ground truths). It can be seen that the parametric approach demonstrates a higher accuracy than F-BFRI. The reason for this is likely related to the fact that the parametric approach has precise control over parameter values. Also, the shapes of the initial fuzzy sets used in F-BFRI are approximated, which may have affected its performance. Nevertheless, both methods seem to have an acceptable level of error when compared with the results in the existing literature [7–9].

The value of υ (the number of discretised intervals per variable) is an important factor for both approaches. Figures 4.5 and 4.6 present the relationships between the approximation errors and the execution times with respect to various values of intervals υ, for the two proposed methods. The results show that the parametric approach produces a higher accuracy when a larger number of intervals is used. However, it is also less scalable. This experimentally demonstrates that the run-time complexity and memory requirement of F-BFRI are more relaxed, which also agrees with the theoretical analysis regarding their respective complexities in Sects. 4.1 and 4.2.

Table 4.5 Errors of P-BFRI and F-BFRI over 500 test samples

$\epsilon\%$	P-BFRI			F-BFRI		
	B	A_3	A_5	B	A_3	A_5
Max	11.32	9.57	8.39	12.70	10.30	10.33
Min	0.0000	0.0073	0.0090	0.0005	0.0004	0.0039
Mean	1.68	3.67	3.12	1.63	5.72	4.95

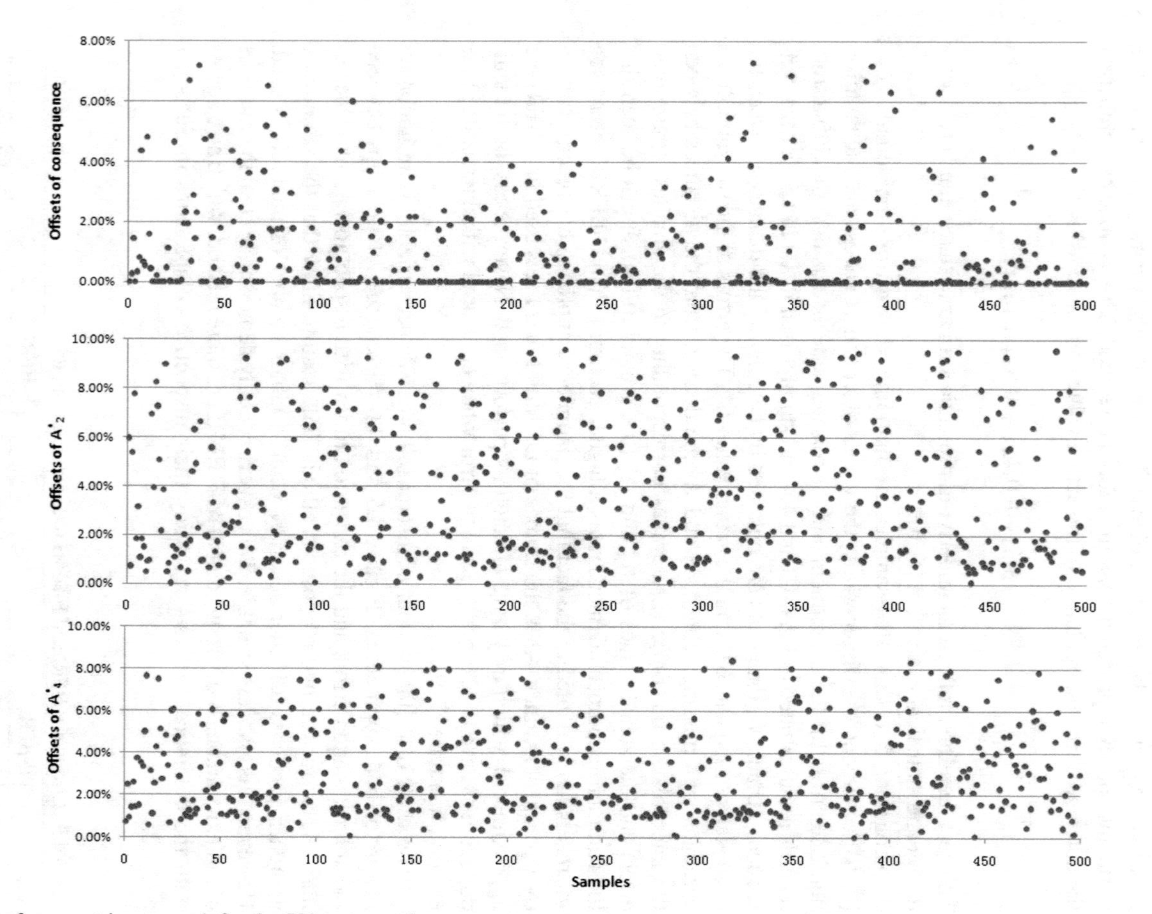

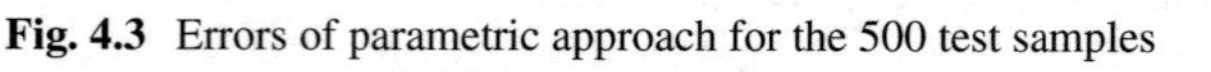

Fig. 4.3 Errors of parametric approach for the 500 test samples

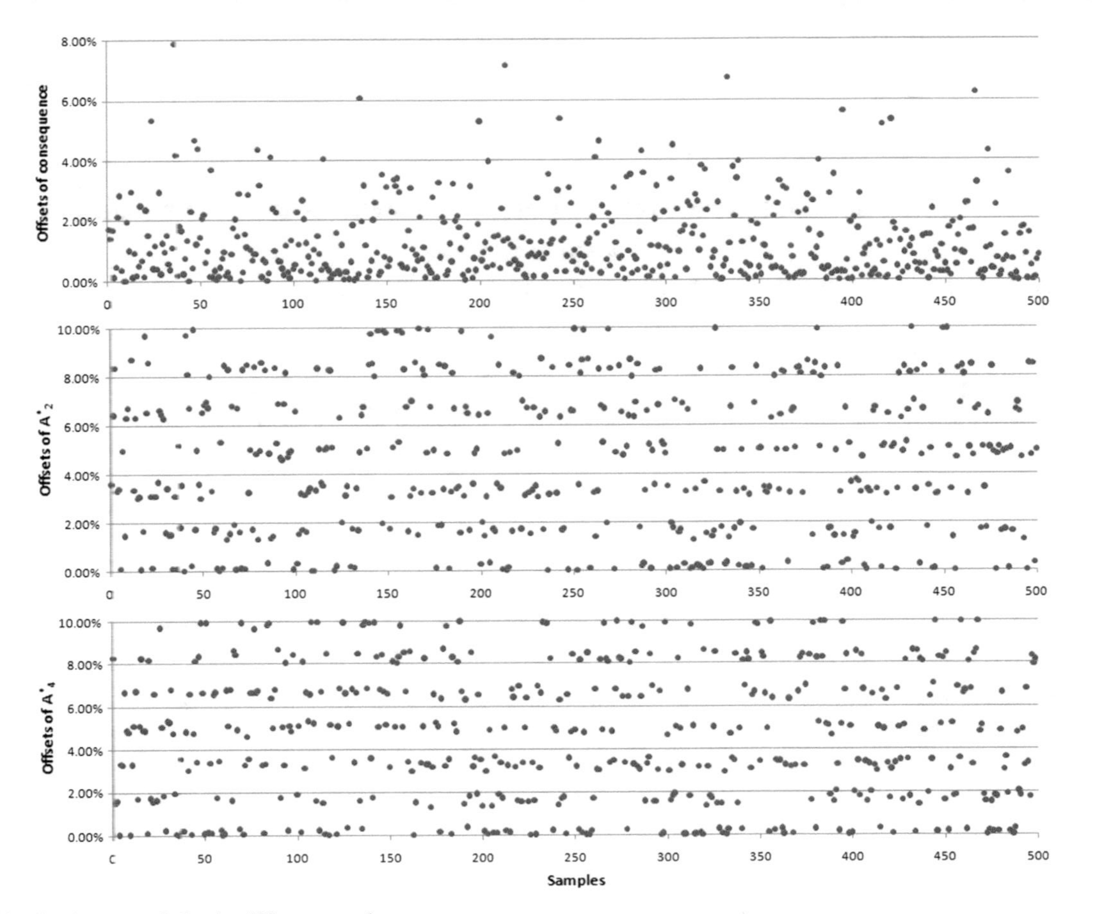

Fig. 4.4 Errors of feedback approach for the 500 test samples

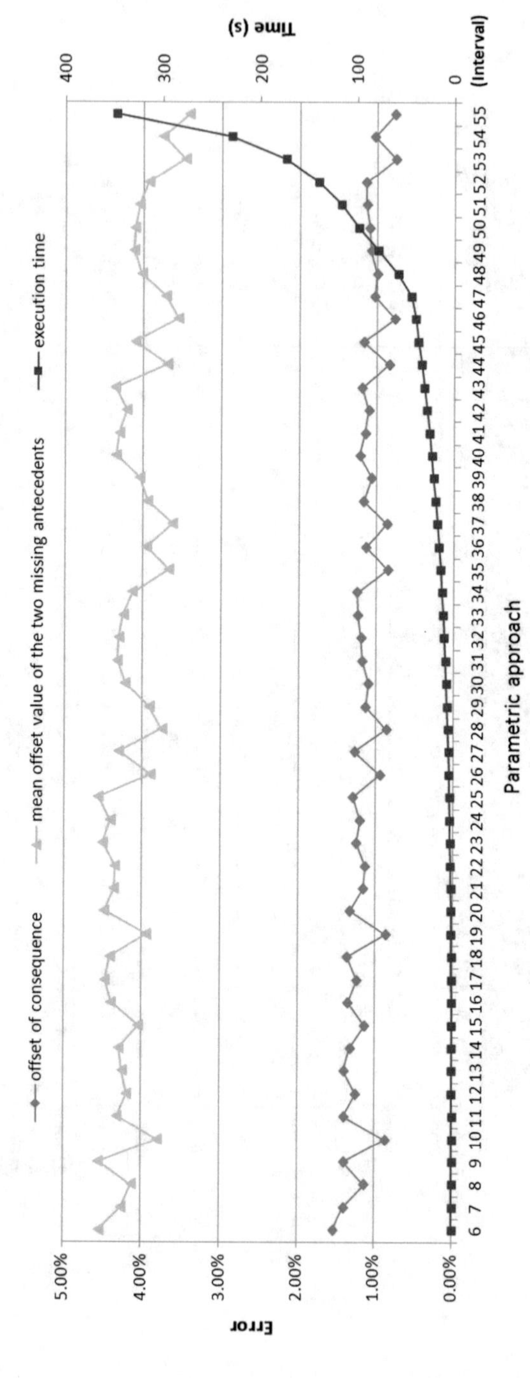

Fig. 4.5 Relationship between the value of υ, approximation error $\epsilon_{\%}$, and execution time for parametric approach

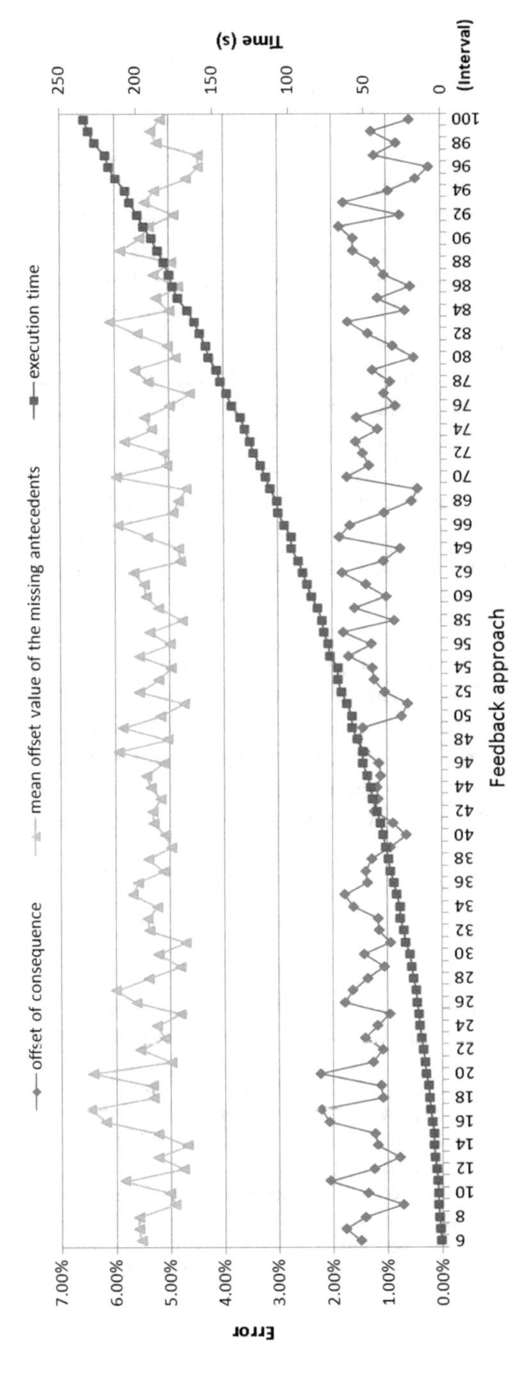

Fig. 4.6 Relationship between the value of υ, approximation error $\epsilon_{\%}$, and execution time for feedback approach

4.4 Summary

This chapter has presented two extensions to BFRI in order to handle multiple missing antecedent values. The proposed techniques are implemented by exploiting the conventional T-FRI mechanism. The parametric approach (P-BFRI) follows the principles of the S-BFRI method, which leads to the closest resemblance of the missing values. The required parameters cannot be obtained in a straightforward manner, due to the lack of sufficient given values. Furthermore, as the number of missing antecedent values increases, the possible scenarios may become extremely wide-reaching or even countless. Fortunately, by definition, all the parameters take values from their underlying ranges. Thus, these ranges are discretised in order to generate the required parameter combinations.

In assessing the performance of the estimation, the conventional T-FRI procedure is then invoked to verify the correctness and accuracy by comparing the output with the actual observed consequent value, so that the most suitable setting is identified. The feedback approach (F-BFRI) is an alternative and more intuitive approach to BFRI. It works by directly estimating the possible initial values of the missing antecedents, then validating the resultant consequent through conventional FRI, in order to identify the most suitable value combination(s) that lead to the observed consequent value. It therefore significantly reduces the time-complexity for parameter estimation.

Worked examples have been provided to illustrate the operation of these approaches. Comparative experimental studies have also demonstrated that the parametric approach is more accurate. However, theoretical study shows that it is computationally more complex. The feedback approach is more intuitive while significantly reducing the time-consuming calculation for parameters, but it is slightly less accurate. Nevertheless, both approaches seem to have an acceptable accuracy.

References

1. S. Jin, R. Diao, Q. Shen, Backward fuzzy interpolation and extrapolation with multiple multi-antecedent rules, in *Proceedings of IEEE International Conference on Fuzzy Systems* (2012), pp. 1170–1177
2. I. Gadaras, L. Mikhailov, An interpretable fuzzy rule-based classification methodology for medical diagnosis. Artif. Intell. Med. **47**(1), 25–41 (2009)
3. A. Tajbakhsh, M. Rahmati, A. Mirzaei, Intrusion detection using fuzzy association rules. Appl. Soft Comput. **9**(2), 462–469 (2009)
4. K.W. Wong, D. Tikk, T.D. Gedeon, L.T. Kóczy, Fuzzy rule interpolation for multidimensional input spaces with applications: a case study. IEEE Trans. Fuzzy Syst. **13**(6), 809–819 (2005)
5. G. Bontempi, H. Bersini, M. Birattari, The local paradigm for modeling and control: from neuro-fuzzy to lazy learning. Fuzzy Sets Syst. **121**(1), 59–72 (2001)
6. L. Kuncheva, Fuzzy versus nonfuzzy in combining classifiers designed by boosting. IEEE Trans. Fuzzy Syst. **11**(6), 729–741 (2003)

7. S.-M. Chen, Y.-C. Chang, Weighted fuzzy rule interpolation based on GA-based weight-learning techniques. IEEE Trans. Fuzzy Syst. **19**(4), 729–744 (2011)
8. Z. Huang, Q. Shen, Fuzzy interpolative reasoning via scale and move transformations. IEEE Trans. Fuzzy Syst. **14**(2), 340–359 (2006)
9. D. Tikk, P. Baranyi, T.D. Gedeon, L. Muresan, Generalization of the rule interpolation method resulting always in acceptable conclusion. Tatra Mt. Math. Publ. **21**, 73–91 (2001)

Chapter 5
An Alternative Backward Fuzzy Rule Interpolation Method

In real-world fuzzy interpolation applications of interconnected rule bases, situations may arise when certain crucial antecedents are absent from given observations. If such missing antecedents were involved in the subsequent interpolation process, the final conclusion would not be deducible using conventional means. To address this issue, an approach termed backward fuzzy rule interpolation and extrapolation has been proposed recently, allowing the observations which directly relate to the conclusion to be inferred or interpolated from the known antecedents and conclusion. As such, it significantly extends the existing fuzzy rule interpolation techniques. Although promising, the existing backward fuzzy rule interpolation concept has only been implemented using an analogical interpolation method [1, 2] as mentioned in the previous chapters. Considering the versatility and potential of the BFRI concept, it is imperative to extend BFRI to also support the α-cut-based methods. In particular, the fuzzy interpolation technique for multidimensional input spaces (IMUL) [3] is employed to extend the existing BFRI approach. This is done because IMUL allows interpolation using the rules with multidimensional input spaces. Also, it guarantees that the interpolative outcomes are crisp if the inputs (observations) are crisp [4, 5].

The rest of this chapter is organised as follows. Section 5.1 presents the proposed methods based on the certain α-cut interpolation method (IMUL), along with several worked examples. In Sect. 5.2, randomised experiments are conducted in order to compare the accuracies and efficiencies of the proposed approach with its alternative (T-FRI-based) implementation. Furthermore, a realistic scenario that concerns terrorist activities is provided as a demonstration of this approach. Section 5.3 summarises the chapter and suggests future enhancements.

5.1 Fuzzy Interpolation for Multidimensional Input Spaces

This section provides an outline of the IMUL method, where the essential concepts are presented. A more detailed explanation, however, can be found in the original work [3]. Trapezoidal fuzzy sets are adopted in the algorithm description in this chapter (triangular fuzzy sets are a special case of the trapezoidal membership function in IMUL

S. Jin et al., *Backward Fuzzy Rule Interpolation*,
https://doi.org/10.1007/978-981-13-1654-8_5

method). For a given rule base $\mathbb{U}$, a fuzzy rule $R \in \mathbb{U}$ with M antecedents $A_i, i = 1, 2, \cdots, M$, and an observation O are expressed in the following format:

R: IF x_1 is $A_1, \cdots$, and x_i is $A_i, \cdots$, and x_M is A_M, THEN y is B

O: x_1 is $A_1^*, \cdots$, and x_i is $A_i^*, \cdots$, and x_M is A_M^*

where $A_i^j = (a_{i,-2}^j, a_{i,-1}^j, a_{i,1}^j, a_{i,2}^j)$ is the ith fuzzy set of the jth rule, which is defined on the domain of the antecedent variable x_i, $i \in \{1, \cdots, M\}$, and M is the total number of antecedents. The observed fuzzy set of variable x_i is denoted by A_i^*.

One of the key notions of IMUL is the reference points $a_{i,0}^j$ of a fuzzy set A_i^j, as illustrated in Fig. 5.1, which is defined as the mid-point of the membership function A_i^j:

$$a_{i,0}^j = \frac{a_{i,-1}^j + a_{i,1}^j}{2} \tag{5.1}$$

The IMUL method typically considers two closest rules $R_p, R_q \in \mathbb{U}$ with regard to a given observation O, which are also illustrated in Fig. 5.1.

For M input dimensions, the reference point b_0^* of the interpolated conclusion B^* is:

$$b_0^* = (1 - \lambda_{\text{core}})b_0^1 + \lambda_{\text{core}} b_0^2 \tag{5.2}$$

where

$$\lambda_{\text{core}} = \frac{\sqrt{\sum_{i=1}^M (a_{i,0}^* - a_{i,0}^1)^2}}{\sqrt{\sum_{i=1}^M (a_{i,0}^2 - a_{i,0}^1)^2}} \tag{5.3}$$

The right core b_1^* of the conclusion is then calculated as:

$$b_1^* = (1 - \lambda_{\text{right}})b_1^1 + \lambda_{\text{right}} b_1^2 + (\lambda_{\text{core}} - \lambda_{\text{right}})(b_0^2 - b_0^1) \tag{5.4}$$

where

$$\lambda_{\text{right}} = \frac{\sqrt{\sum_{i=1}^M (a_{i,1}^* - a_{i,1}^1)^2}}{\sqrt{\sum_{i=1}^M (a_{i,1}^2 - a_{i,1}^1)^2}} \tag{5.5}$$

The left core b_{-1}^* is obtained in the same manner.

After calculating the two sides of the core, the two flanks can also be determined, where the relative fuzziness of the fuzzy sets in all of the input dimensions is taken into consideration. An illustration of the necessary parameters involved in this step is given in Fig. 5.2. The procedures for calculating the right flank of the conclusion are provided below.

Based on A_i^2 and B^2, the fuzziness of the antecedents s_i and consequent s' can be calculated as:

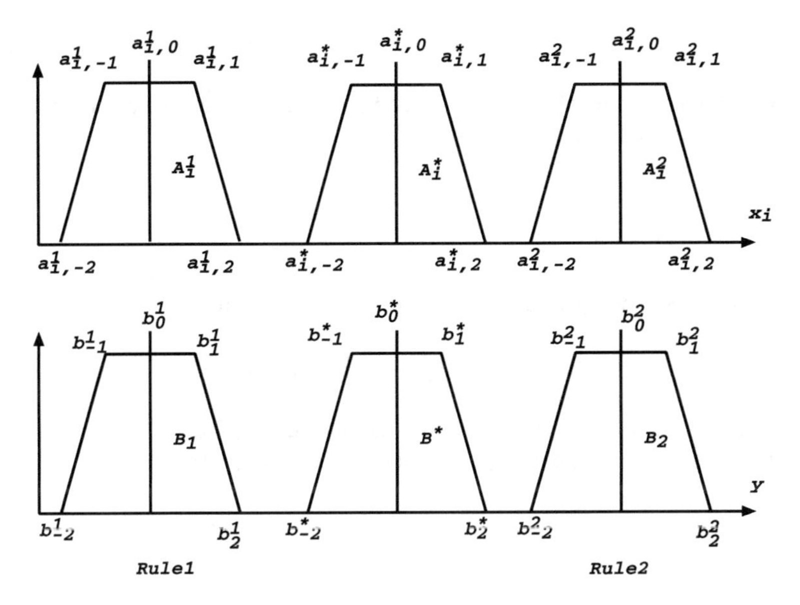

Fig. 5.1 Notations used for IMUL method

Fig. 5.2 Ratio for the fuzziness in IMUL method

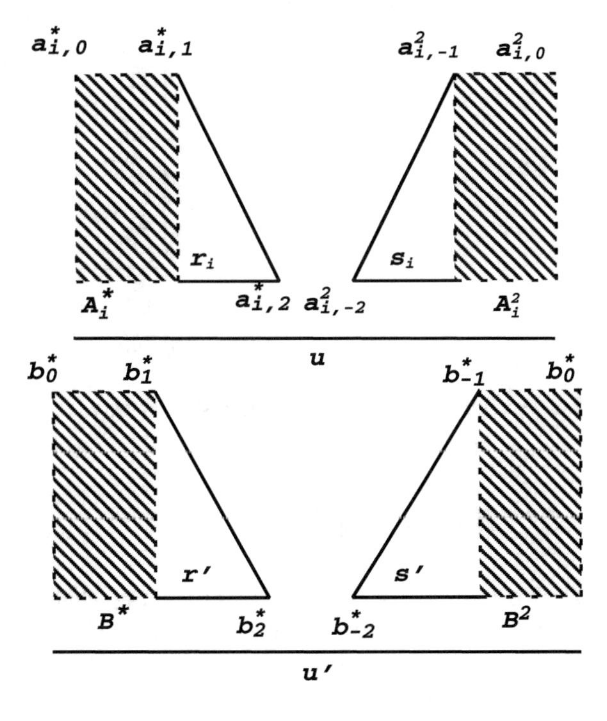

$$s_i = a_{i,-2}^2 - a_{i,-1}^2, \quad s' = b_{-2}^2 - b_{-1}^2 \tag{5.6}$$

and the fuzziness of the observed antecedents r_i and the interpolated consequent r' can then be obtained as follows:

$$r_i = a_{i,2}^* - a_{i,1}^* \tag{5.7}$$

$$r' = b_2^* - b_1^* \tag{5.8}$$

The distances between the observation and the right side rule are calculated as follows:

$$u_i = a_{i,0}^2 - a_{i,0}^*, \quad u' = b_0^2 - b_0^* \tag{5.9}$$

The right flank b_2^* of the interpolated consequent fuzzy set can then be computed, using the sum of the parameters fuzziness r' and the right core b_1^*:

$$b_2^* = b_1^* + r', r' = r\left(1 + |\frac{s'}{u'} - \frac{s}{u}|\right) \tag{5.10}$$

Here r' is a bias ratio which is aggregated from the corresponding values of s_i, r_i and u_i. For multidimensional inputs, the process of aggregation is given as follows:

$$s = \sqrt{\sum_{i=1}^{M} (s_i)^2} \tag{5.11}$$

$$r = \sqrt{\sum_{i=1}^{M} (r_i)^2} \tag{5.12}$$

$$u = \sqrt{\sum_{i=1}^{M} (u_i)^2} \tag{5.13}$$

The left flank b_{-2}^* is calculated in a similar way.

5.1.1 BFRI Using IMUL

The basic concept of BFRI involves situations where the consequent value is known, and the values of all but one antecedent variable are also given. The task is to estimate the value of the unknown antecedent. Therefore, the BFRI concept can then be defined in the following form:

$$A_l^* = f_{\text{BFRI}}((B^*, A_1^*, \cdots, A_{l-1}^*, A_{l+1}^*, \cdots, A_M^*), (R_1, R_2)) \tag{5.14}$$

where f_{BFRI} denotes the entire process of obtaining A_l^*, the unknown (or required) antecedent value, which is to be backward interpolated. It uses the two closest rules, with regard to the observed (or predicted) values from the $(M-1)$ antecedents and the conclusion B^*.

A close examination of the IMUL algorithm reveals that, in order to successfully backward interpolate the missing value, the two closest rules need to initially be identified. All of the parameters involved in IMUL (for trapezoidal fuzzy sets): λ_{core}, λ_{right}, λ_{left}, s, r and u also need to be computed for the known antecedent variables, and the now observed consequent variable. The acquisition of these essential parameters allows a possible calculation process to be derived, which then helps to restore the missing antecedent value. The proposed IMUL-based implementation of the BFRI idea (IMUL-BFRI) that reflects this intuition is summarised below.

5.1.1.1 Determination of the Closest Rules

In reference to the earlier definition of the BFRI process in Eq. 5.14, when $B^*, (A_1^*, \cdots, A_{l-1}^*, A_{l+1}^*, \cdots, A_M^*)$ are given, in order to interpolate/extrapolate the unknown antecedent A_l^*, the discovery of the two closest rules R_1, R_2 are required. The distance d between a rule and an observation is determined by computing the aggregated distance of all the antecedent variables:

$$d = \sqrt{\sum_{i=1}^{M} d(A_i, A_i^*)^2} \tag{5.15}$$

where

$$d(A_i, A_i^*) = \frac{d(a_{i,0}, a_{i,0}^*)}{range_{x_i}} \tag{5.16}$$

where $range_{x_i} = sup\{x_i\} - inf\{x_i\}$ is the domain range of the variable x_i, and $d(A_i, A_i^*) \in [0, 1]$ is the normalised result of the otherwise absolute distance measure. This is to ensure that the distances are compatible with each other over different variable domains. The two rules which have the least distance measurements with regard to the observed values A_i^* and the conclusion B^* are then chosen to be used in the later steps. Instead of using the distance measure method, a modified scheme may also be employed which better reflects the biased consideration towards the consequent variable (as per the intuition indicated above):

$$d = \sqrt{{d_B}^2 + \sum_{i=1,\ i \neq l}^{M} {d_{A_i}}^2} \tag{5.17}$$

5.1.1.2 BFRI with Multi-antecedent Rules

To aid the explanation, assume that the observation O and two closest rules R_1, R_2 that are returned by this step are represented as follows:

O: x_1 is $A_1^*, \cdots, x_{l-1}$ is $A_{l-1}^*, \cdots, x_{l+1}$ is $A_{l+1}^*, \cdots, x_M$ is A_M^*
R_1: IF x_1 is $A_1^1, \cdots,$ and x_i is $A_i^1, \cdots,$ and x_M is A_M^1, THEN y is B^1
R_2: IF x_1 is $A_1^2, \cdots,$ and x_i is $A_i^2, \cdots,$ and x_M is A_M^2, THEN y is B^2

Intuitively, the parameter for the missing antecedent should be calculated by subtracting parameter values of the known antecedents from the conclusion. Therefore, the first step of the actual interpolation process is to create the reference point $a_{l,0}^*$ of the missing fuzzy set A_l^*. For this, λ_{coreB} needs to be calculated by modifying Eq. 5.2, such that:

$$\lambda_{coreB} = \frac{b_0^* - b_0^1}{b_0^2 - b_0^1} \tag{5.18}$$

Recall that $a_{l,0}^*$ is part of the λ_{coreB} on the right-hand side of Eq. 5.3. Now that B^* is already known, λ_{coreB} along with $\lambda_{coreA_i^j}$ for the other known antecedents A_i^j $(i = 1, \ldots, M, i \neq l)$ can be calculated as:

$$a_{l,0}^* = a_{l,0}^1 + \sqrt{\left| M(\lambda_{coreB})^2(b_0^* - b_0^1)^2 - \sum_{i=1, i\neq l}^{M} (a_{i,0}^* - a_{i,0}^1)^2 \right|} \tag{5.19}$$

Having the reference point $a_{l,0}^*$, by reversing the forward interpolation procedure, the right core $a_{l,1}^*$ can now be computed:

$$a_{l,1}^* = a_{l,1}^1 + \sqrt{\left| M(\lambda_{rightB})^2(b_1^* - b_1^1)^2 - \sum_{i=1, i\neq l}^{M} (a_{i,1}^* - a_{i,1}^1)^2 \right|} \tag{5.20}$$

where

$$\lambda_{rightB} = \frac{b_1^* - b_1^1}{b_1^2 - b_1^1} \tag{5.21}$$

Following the same reasoning steps as those of the right core $a_{l,1}^*$, the left core $a_{l,-1}^*$ can also be calculated:

$$a_{l,-1}^* = a_{l,-1}^1 + \sqrt{\left| M(\lambda_{leftB})^2(b_{-1}^* - b_{-1}^1)^2 - \sum_{i=1, i\neq l}^{M} (a_{i,-1}^* - a_{i,-1}^1)^2 \right|} \tag{5.22}$$

where

$$\lambda_{\text{left}B} = \frac{b^{*}_{-1} - b^{1}_{-1}}{b^{2}_{-1} - b^{1}_{-1}} \tag{5.23}$$

Having obtained the two sides of the core $a^{*}_{l,1}$ and $a^{*}_{l,-1}$, according to Eq. 5.7, the right flank of the missing value $a^{*}_{l,2}$ is calculated as follows:

$$a^{*}_{l,2} = r_l + a^{*}_{l,1} \tag{5.24}$$

Considering the fuzzy sets in all the known antecedent and conclusion spaces, the relative fuzziness parameter r_l can be calculated by modifying Eq. 5.12, such that:

$$r_l = \sqrt{\left| r^{2}_{B^{*}} - \sum_{i=1, i \neq l}^{M} (r_i)^2 \right|} \tag{5.25}$$

where r_i and $r_{B^{*}}$ are obtained according to Eqs. 5.7 and 5.8, respectively.

$$r_{B^{*}} = b^{*}_{2} - b^{*}_{1} \tag{5.26}$$

The left flank $a^{*}_{l,-2}$ is obtained analogously:

$$\begin{cases} a^{*}_{l,-2} = a^{*}_{l,-1} - r_l \\ r'_l = \sqrt{(r'_{B^{*}})^2 - \sum_{i=1, i \neq l}^{M} (r_i)^2} \\ r'_{B^{*}} = b^{*}_{-1} - b^{*}_{-2} \end{cases} \tag{5.27}$$

Finally, with all points $a^{*}_{l,-2}$, $a^{*}_{l,-1}$, $a^{*}_{l,1}$, $a^{*}_{l,2}$ calculated, the backward interpolated value A^{*}_{l} can be assembled.

5.1.2 Worked Examples

This section provides two worked examples of the proposed IMUL-BFRI approach. For each of these, the value of the consequent variable is obtained using randomly chosen values for the antecedent variables. The "missing" value is then (purposefully) removed from the observation, allowing the application of BFRI.

Example 5.1.1 IMUL-BFRI with Trapezoidal Fuzzy Sets

This example illustrates IMUL-BFRI involving multiple antecedent rules, where the variable values are represented by trapezoidal membership functions. The observation and the two closest rules are given in Table 5.1 and Fig. 5.3. Here, A^{*}_{3} is the missing antecedent which is to be inferred.

Table 5.1 Two closest rules for observation in Example 5.1.1

	O	R_1	R_2
x_1	(3.5, 4.0, 5.0, 7.0)	(0.2, 1.1, 2.2, 2.7)	(10.5, 11.5, 12.5, 13.1)
x_2	(5.0, 5.5, 6.0, 7.5)	(1.5, 2.0, 2.5, 3.0)	(10.0, 11.2, 12.3, 13.0)
x_3	missing	(0.4, 1.5, 2.0, 2.5)	(10.2, 11.0, 11.5, 13.2)
x_4	(4.5, 5.2, 6.5, 7.5)	(1.1, 1.5, 2.1, 2.5)	(10.1, 12.0, 12.5, 14.3)
y	(5.5, 6.5, 7.0, 8.7)	(0.2, 2.0, 2.5, 3.0)	(12.0, 13.0, 13.5, 14.2)

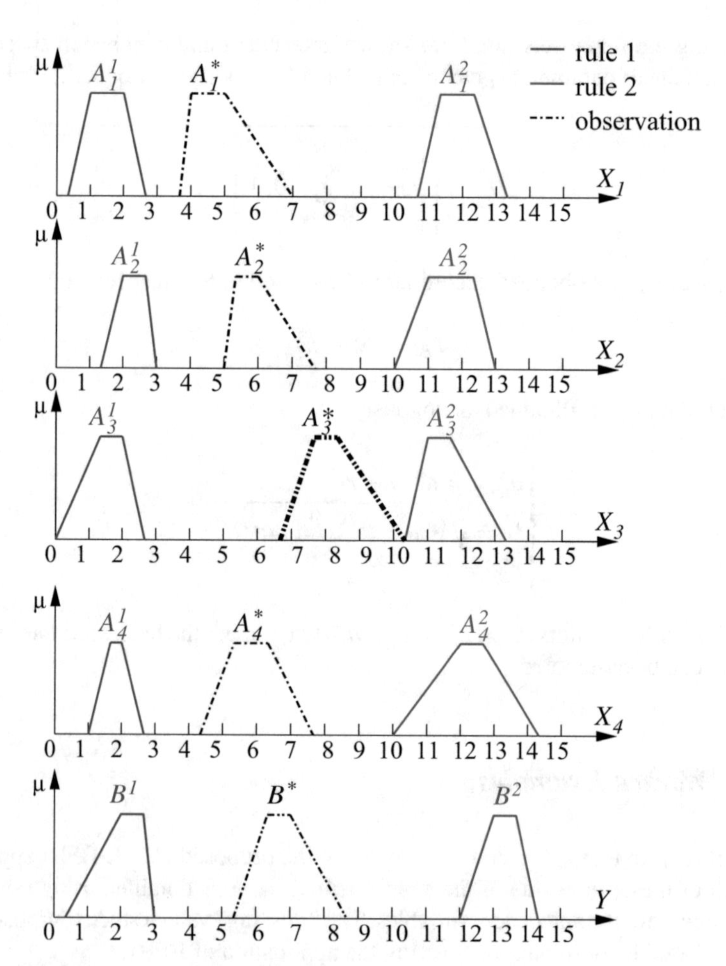

Fig. 5.3 Example of IMUL-BFRI with multi-antecedent (trapezoidal set) rules

1. *Backward Interpolation*:
   In order to successfully backward interpolate the missing value A_3^*, the reference point of A_3^*: $a_{3,0}^*$ needs to be obtained first. According to Eq. 5.18, λ_{core} of B^* can

be obtained $\lambda_{coreB} = 0.409$. According to Eq. 5.19, the reference point $a^*_{3,0} = 7.56$. The right core and the left core of $A^*_3 : a^*_{3,1}$, and $a^*_{3,-1}$ can be derived from Eqs. 5.20 and 5.21, which are 8.04 and 7.37. According to Eq. 5.21, $\lambda_{rightB} = 0.440$ and $\lambda_{leftB} = 0.327$ According to Eqs. 5.24–5.27, $r_{B^*} = 0.864$, $r_3 = 0.492$, $r'_{B^*} = 1.700$, $r'_3 = 2.088$, the right flank and the left flank of the missing value A^*_3: $a^*_{3,2}$ and $a^*_{3,-2}$ are 10.12 and 6.88. Finally, the backward interpolative conclusion $A^*_3 = (6.88, 7.37, 8.04, 10.12)$.

2. *Verification*:
The result of BFRI can be verified by performing the conventional IMUL method, using the reconstructed observation involving A^*_3. Applying forward interpolation results in the conclusion $B^* = (5.36, 6.69, 7.19, 10.97)$, $b^*_0 = 6.94$. This is consistent with the given observed conclusion (5.50, 6.50, 7.00, 8.70), which has a representative value of 6.98 (Table 5.1).

Example 5.1.2 IMUL-BFRI with Triangular Fuzzy Sets and Singleton Values.

To further demonstrate the generality of the proposed approach, this example illustrates IMUL-BFRI involving triangular membership functions and also singleton values. The two adjacent rules which involve singleton fuzzy sets are given in Table 5.2 and Fig. 5.4, with the observation being $A^*_1 = (4, 5, 6)$, $A^*_2 = (5, 6, 7)$, $B^* = (10, 11, 13)$.

1. *Backward Interpolation*:
In order to successfully backward interpolate the missing value A^*_3, the reference point of A^*_3: $a^*_{3,0}$ needs to be obtained first. According to Eq. 5.18, λ_{core} of B^* can be obtained $\lambda_{coreB} = 0.40$. According to Eq. 5.19, the reference point $a^*_{3,0} = 8.01$. The right core and the left core of $A^*_3 : a^*_{3,1}$, and $a^*_{3,-1}$ can be derived from Eqs. 5.20 and 5.21, which are 8.01 and 7.95. According to Eq. 5.21, $\lambda_{rightB} = 0.40$ and $\lambda_{leftB} = 0.40$ According to Eqs. 5.24–5.27, $r_{B^*} = 2.0$, $r_3 = 1.41$, $r'_{B^*} = 0.38$, $r'_3 = 1.36$, the right flank and the left flank of the missing value A^*_3: $a^*_{3,2}$ and $a^*_{3,-2}$ are 9.42 and 6.59. Finally, the backward interpolative conclusion $A^*_3 = (6.59, 7.97, 9.42)$.
2. *Verification*:
The result can be validated here by performing conventional IMUL using the obtained A^*_3, resulting in the conclusion being (8.89, 11.0, 13.27), which is within an acceptable degree of error with regard to the given observed conclusion (10, 11, 13).

Table 5.2 Two closest rules for observation in Example 5.1.2

	O	R_1	R_2
x_1	(4, 5, 6)	(2, 2, 2)	(7, 9, 10)
x_2	(5, 6, 7)	(3, 3, 3)	(8, 9, 10)
x_3	missing	(4, 4, 4)	(9, 10, 11)
y	(10, 11, 13)	(7, 7, 7)	(15, 17, 19)

5.2 Experimentation and Discussion

5.2.1 *Synthetic Evaluation*

To systematically compare the proposed IMUL-BFRI method with the original implementation using scale and move transformation-based interpolation method (T-FRI-BFRI), a numerical function shown in Eq. 5.28 is used. The rule base employed in the experimental evaluation is generated using the following steps: (1) a random set of crisp values are selected for the function variables, and the outcome is calculated according to Eq. 5.28; (2) these crisp values are then fuzzified into trapezoidal fuzzy sets; and (3) the rule base is then populated using these randomly generated rules, while checking (and where appropriate, removing) rules to ensure the underlying domain is reasonably covered, while there still exist sufficient gaps between rules in order to encourage interpolation.

$$y = 3x_1 - 3.3x_2 + 0.4x_3 + 0.5x_4 + 0.7x_5 - 2.0x_6 \\ +1.5x_7 - 3.1x_8 + 0.3x_9 + 1.2x_{10} \tag{5.28}$$

This experimental set up enables an initial sparse rule base to be generated, that is, an approximation of the underlying knowledge, simulating those obtainable by "subject experts". An observation is obtained in a similar manner, where the "missing" values are then purposefully removed to facilitate backward reasoning. Since, the underlying function, i.e., "ground truth", is available. The consistency, accuracy and robustness of the interpolative procedure can then be verified, by comparing the outcome of the interpolation to the actual value computed using Eq. 5.28. This test, therefore, reflects an underlying principle similar to that behind cross-validation and statistical evaluation [6, 7].

The first set of experiments focuses on accuracy evaluation, where the first five input variables $x_i, i = 1, \ldots, 5$, are considered. Altogether, 500 simulated samples (observations) are randomly drawn from the domain $U = [0, 10]^5$. Without loss of generality, each antecedent value in the observation is assumed to be missing. The errors for the results of forward interpolation and the backward interpolation are summarised in Table 5.3. The error is measured here with percentage error $\epsilon_{\%}$:

$$\epsilon_{\%} = 100 \times \frac{b_0^j - b_0^{j*}}{sup\{B\} - inf\{B\}} \tag{5.29}$$

For the antecedent A_i and the consequent variable B, the errors are obtained by calculating the distance between the backward inferred antecedent A_i^j and forward inferred consequent B^j, $(i = 1, 2, \ldots, 5, j = 1, 2, \ldots, 500)$ and the actual value A_i^{j*} and B^{j*}, $(i = 1, 2, \ldots, 5, j = 1, 2, \ldots, 500)$. The errors of the antecedent variables are derived from the distances between the interpolative outcomes and the actual values (the ground truths). It can be seen that the T-FRI-based approach demonstrates a

Table 5.3 Errors of the IMUL-BFRI and T-FRI-BFRI over 500 test samples

	$\epsilon\%$			
	IMUL-BFRI		T-FRI-BFRI	
	B	$A_i, (i = 1, \ldots, 5)$	B	$A_i, (i = 1, \ldots, 5)$
S.D	1.21	0.81	1.03	0.74
Mean	3.41	6.98	3.21	6.07

Table 5.4 Example rules for Explosion likelihood

	Crowdedness	Warning	Explosion likelihood
C1	Very Low	Moderate	Very Low
C2	Moderate	Moderate	Low
C3	Moderate	High	Low
C4	High	Moderate	Moderate

little higher accuracy than the IMUL-based approach, in terms of both forward and backward methods. Nevertheless, both methods seem to have an acceptable level of error when compared with the results in the existing literature [1, 8, 9]. Moreover, this result does demonstrate that the general concept of BFRI can be potentially achieved via interpolation methods of any type.

The second part of the evaluation aims to compare the execution times of IMUL-BFRI and T-FRI-BFRI. For this, rule bases with varying numbers of antecedents (from 1 to 10) are generated based on Eq. 5.28. Figure 5.5 presents a comparison of execution times with respect to varying numbers of input antecedents. The result shows that the run-time efficiency of IMUL-BFRI is much more competitive than T-FRI-BFRIand is more scalable for more complex (with a larger number of antecedents in the rule bases).

5.2.2 Practical Scenario

In this section, a practical scenario that involves the possible detection of a terrorist bombing threat is considered; an illustration of this scenario is given in Fig. 5.6. The likelihood of an explosion can be directly affected by the number of people in the area; a crowded place is usually more likely to attract the attention of terrorists. The number of public warning signs displayed in the area may also affect the potential outcome. With many eye-catching warning signs, people may raise their awareness of the surroundings and may promptly report suspicious individuals or items, giving less opportunities for the terrorists to attack. Given such concepts regarding the antecedent variables *Crowdedness*, *Warning* level and the consequent variable *Explosion Likelihood*, a rule base can be established with example rules listed in Table 5.4.

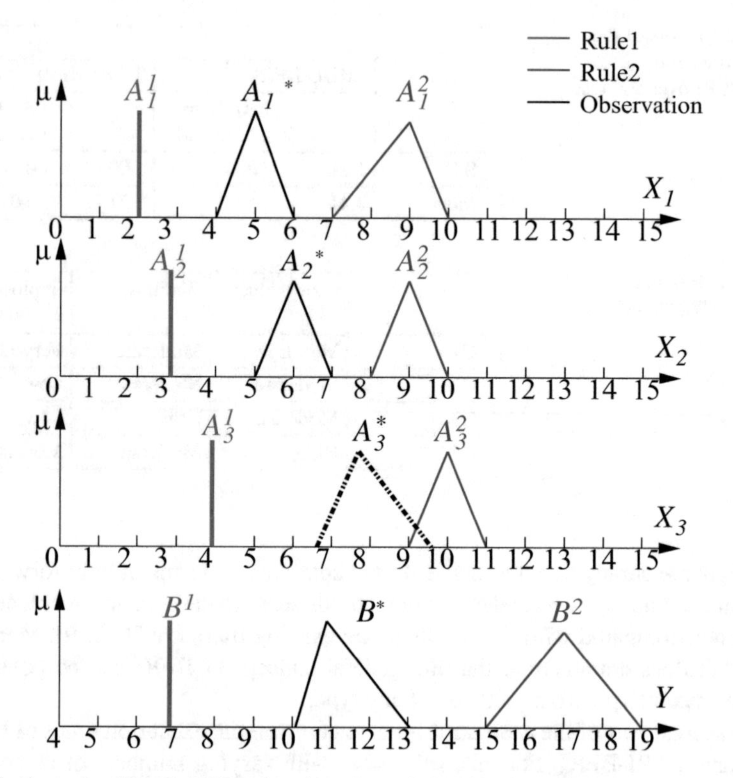

Fig. 5.4 Example of IMUL-BFRI with triangular and singleton fuzzy sets

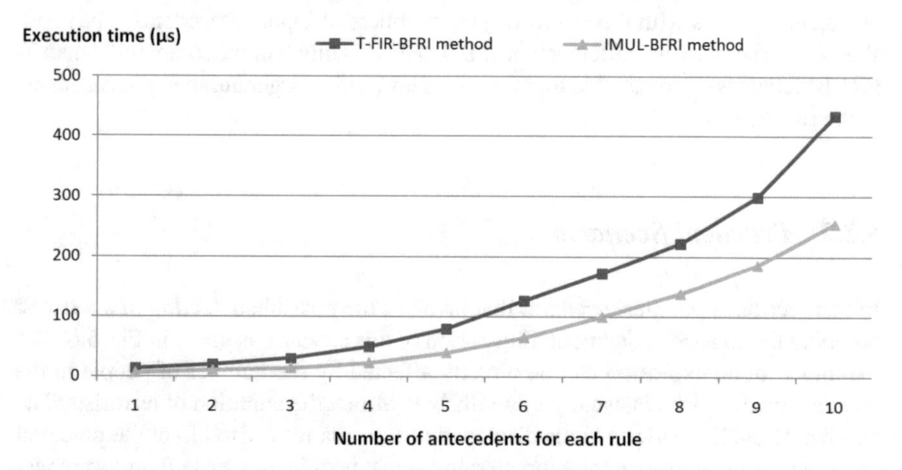

Fig. 5.5 Relationship between the number of antecedents and execution time

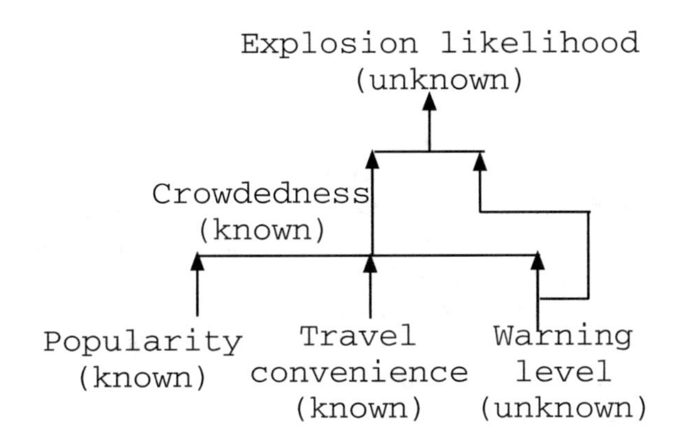

Fig. 5.6 Hierarchical fuzzy reasoning structure for terrorist bomb threat

Additionally, there may exist another rule base that focuses on the prediction of crowdedness. The number of people in an area is directly related to the *Popularity* of the location, but it can also be affected by the level of *Travel* convenience. A well-known place that is hard to get to by usual means may not have more people than an attraction that is less famous, but very easy to reach. Moreover, considering the current scenario, the crowdedness may also change in relation to the *Warning* level. The less brave individuals may shy away from locations that are considered dangerous, judged by the amount of explosion alerts in the surrounding areas. Summarising the above information, the second rule base may be derived. Table 5.5 displays a subset of such rules.

It is easy to identify that both rule bases involved are fuzzy and sparse. Fuzziness is naturally introduced by the presence of linguistic terms used to describe the domain variables. From the linguistic terms, it can be deduced that the rule bases contain substantial gaps amongst the underlying domains of the variables concerned. The terms *Low*, *Moderate* and *High* although provide reasonable coverage of the entire domain, but intermediate values such as *Moderate Low*, *Moderate High* are not

Table 5.5 Example rules for crowdedness

	Popularity	Travel	Warning	Crowdedness
C1	Very Low	Very Low	Very high	Very Low
C2	Very Low	Very high	Very Low	Low
C3	Very Low	High	High	Low
C4	Moderate	Moderate	High	Low
C5	Moderate	High	Low	Moderate
C6	High	Very Low	High	Very Low
C7	High	High	Moderate	Moderate
C8	High	High	Very Low	High

represented. By converting the linguistic terms into fuzzy sets, the following two rule bases presented in Tables 5.6 and 5.7 may be produced.

For traditional forward interpolative reasoning, in order to interpolate *Explosion likelihood*, the observed value for the level of *Crowdedness* and *Warning* level are both required. The antecedent variable *Warning* level is particularly important for it is required by both rule bases. Without it, no matter what other information is available, even when the *Crowdedness* is known, as illustrated in Table 5.8, forward interpolation would still fail.

In order to interpolate *Explosion likelihood*, it is essential to determine the value of *Warning* using BFRI. Following the steps detailed in Sect. 5.1.1.1, the two closest rules are first selected using the consequence-biased distance measure stated in Eq. 5.17, as shown in Table 5.9. By producing the backward interpolated value $A^*_{warning} = (5.2, 6.4, 7.4)$. Now with both antecedents $A^*_{crowdedness} = (4.1, 5.1, 6.1)$ and $A^*_{warning} = (5.2, 6.4, 7.4)$ present, the second rule base can therefore be evoked to produce the final interpolation result, $B^*_{explosion} = (2.4, 4.3, 6.7)$, this time using the standard IMUL method.

If the normal distance measure in Eq. 5.15 is used, the rules will no longer be selected with biased focus on the consequence, instead of treating everything with equal weight. The rules shown in Table 5.10 will then be selected. Following the backward interpolation process again, a different outcome of $A^*_{warning} = (3.0, 4.2, 5.2)$ will be derived, followed by $B^*_{explosion} = (0.8, 2.7, 4.5)$ interpolated using forward IMUL on the second rule base. Looking back at the original observation, given the two known antecedent values $A^*_{popularity} = (5.5, 6.5, 7.5)$ and $A^*_{travel} = (5.4, 6.4, 7.4)$, the intuitive deduction of *Crowdedness* should be quite high, as the location is both moderately high in popularity and is moderately convenient to reach. The only reason that the observed *Crowdedness* has a moderate value of (4.1, 5.1, 6.1) may well have been caused by a reasonable level of *Warning*. Intuitively, the outcome

Table 5.6 Fuzzy rule base for Explosion likelihood

	Crowdedness	Warning	Explosion likelihood
E1	(0.0,1.0,2.0)	(0.0,0.8,1.8)	(0.0,0.9,1.9)
E2	(0.0,1.0,2.0)	(3.4,4.4,5.4)	(0.0,0.7,1.7)
E3	(0.1,1.1,2.1)	(7.4,8.4,9.4)	(0.0,0.6,1.6)
E4	(3.5,4.5,5.5)	(0.0,0.5,1.5)	(3.2,4.2,5.2)
E5	(3.7,4.7,5.7)	(3.1,4.1,5.1)	(2.3,3.3,4.3)
E6	(4.2,5.2,6.2)	(6.9,7.9,8.9)	(1.9,2.9,3.9)
E7	(7.6,8.6,9.6)	(0.0,0.3,1.3)	(7.2,8.2,9.2)
E8	(7.0,8.0,9.0)	(3.5,4.5,5.5)	(4.5,5.5,6.5)
E9	(7.7,8.7,9.7)	(7.6,8.6,9.6)	(3.7,4.7,5.7)

Table 5.7 Fuzzy rule base for crowdedness

	Popularity	Travel	Warning	Crowdedness
C1	(0.0,0.4,1.4)	(0.0,0.5,1.5)	(0.0,0.9,1.9)	(0.0,0.8,1.8)
C2	(0.2,1.2,2.2)	(0.0,1.0,2.0)	(7.3,8.3,9.3)	(0.0,0.7,1.7)
C3	(0.0,0.4,1.4)	(3.0,4.0,5.0)	(0.0,0.9,1.9)	(0.8,1.8,2.8)
C4	(0.0,0.6,1.6)	(3.6,4.6,5.6)	(3.2,4.2,5.2)	(0.5,1.5,2.5)
C5	(0.0,0.7,1.7)	(7.6,8.6,9.6)	(0.0,1.0,2.0)	(1.8,2.8,3.8)
C6	(0.0,0.8,1.8)	(7.1,8.1,9.1)	(6.9,7.9,8.9)	(0.7,1.7,2.7)
C7	(3.2,4.2,5.2)	(0.0,1.0,2.0)	(0.0,1.0,2.0)	(1.1,2.1,3.1)
C8	(3.5,4.5,5.5)	(0.0,0.9,1.9)	(7.3,8.3,9.3)	(0.3,1.3,2.3)
C9	(3.4,4.4,5.4)	(3.2,4.2,5.2)	(3.9,4.9,5.9)	(2.0,3.0,4.0)
C10	(3.1,4.1,5.1)	(7.4,8.4,9.4)	(0.0,0.9,1.9)	(4.5,5.5,6.5)
C11	(3.2,4.2,5.2)	(7.7,8.7,9.7)	(6.8,7.8,8.8)	(2.5,3.5,4.5)
C12	(7.6,8.6,9.6)	(0.2,1.2,2.2)	(0.0,0.5,1.5)	(2.3,3.3,4.3)
C13	(7.4,8.4,9.4)	(0.1,1.1,2.1)	(3.0,4.0,5.0)	(1.4,2.4,3.4)
C14	(6.8,7.8,8.8)	(0.0,0.5,1.5)	(7.4,8.4,9.4)	(0.5,1.5,2.5)
C15	(7.2,8.2,9.2)	(3.4,4.4,5.4)	(3.6,4.6,5.6)	(3.2,4.2,5.2)
C16	(7.3,8.3,9.3)	(7.2,8.2,9.2)	(0.0,0.6,1.6)	(6.7,7.7,8.7)
C17	(7.6,8.6,9.6)	(7.0,8.0,9.0)	(7.1,8.1,9.1)	(3.6,4.6,5.6)

Table 5.8 Observation

Popularity	Travel	Warning	Crowdedness
(5.5,6.5,7.5)	(5.4,6.4,7.4)	N/A	(4.1, 5.1, 6.1)
Moderate High	Moderate High	N/A	Moderate

Table 5.9 Two closest rules using biased distance measure

Popularity	Travel	Warning	Crowdedness
(7.4,8.4,9.4)	(6.8,7.8,8.8)	(3.4,4.4,5.4)	(4.7,5.7,6.7)
(3.4,4.4,5.4)	(3.2,4.2,5.2)	(3.9,4.9,5.9)	(2.0,3.0,4.0)

Table 5.10 Two closest rules using plain distance measure

Popularity	Travel	Warning	Crowdedness
(7.2,8.2,9.2)	(3.4,4.4,5.4)	(3.6,4.6,5.6)	(3.2,4.2,5.2)
(0.0,0.6,1.6)	(3.6,4.6,5.6)	(3.2,4.2,5.2)	(0.5,1.5,2.5)

$A^*_{warning} = (5.2, 6.4, 7.4)$ from the biased distance measure is therefore more agreeable than $A^*_{warning} = (3.0, 4.2, 5.2)$ from the plain distance measure. Further experiments show that, by using $A^*_{warning} = (3.0, 4.2, 5.2)$ and the two known antecedent values, forward IMUL method interpolates *Crowdedness* as $(1.8, 3.2, 5.5)$, which is much further than the original observation.

5.3 Summary

This chapter has presented a new BFRI method that complements IMUL by supporting backward inference, which allows flexible interpolation when certain antecedents are missing from the observation. The work is based upon the α-cut interpolation mechanisms rather than T-FRI-based methods (Chaps. 3 and 4), in order to demonstrate the flexibility of BFRI approach in dealing with two groups of interpolation methods. Instead of using the distance metric method, a modified scheme may also be employed which better reflects the biased consideration towards the consequent variable. The parameters for the missing antecedent are calculated by subtracting parameter values of the known antecedents from the conclusion.

Worked examples are provided to illustrate their operation. The results show that the proposed method in this chapter can handle multiple multi-antecedent rules and ensure the maintenance of convexity and normality of interpolated outcomes. Synthetic experiments and a practical application are supplied in order to showcase the effectiveness of the proposed approach in addressing real-world problems.

References

1. Z. Huang, Q. Shen, Fuzzy interpolative reasoning via scale and move transformations. IEEE Trans. Fuzzy Syst. **14**(2), 340–359 (2006)
2. Z. Huang, Q. Shen, Fuzzy interpolation and extrapolation: a practical approach. IEEE Trans. Fuzzy Syst. **16**(1), 13–28 (2008)
3. K.W. Wong, D. Tikk, T.D. Gedeon, L.T. Kóczy, Fuzzy rule interpolation for multidimensional input spaces with applications: a case study. IEEE Trans. Fuzzy Syst. **13**(6), 809–819 (2005)
4. A. Gupta, H. Eren, Mathematical modeling and on-line control of hydrocyclones. Proc. Aus. IMM **295**(2), 31–41 (1990)
5. M.H. Rider, *The Geological Interpretation of Well Logs*, 2nd edn. (Whittles, Caithness, Scotland, 1996)
6. G. Bontempi, H. Bersini, M. Birattari, The local paradigm for modeling and control: from neuro-fuzzy to lazy learning. Fuzzy Sets Syst. **121**(1), 59–72 (2001)
7. L. Kuncheva, Fuzzy versus nonfuzzy in combining classifiers designed by boosting. IEEE Trans. Fuzzy Syst. **11**(6), 729–741 (2003)
8. S.-M. Chen, Y.-C. Chang, Weighted fuzzy rule interpolation based on GA-based weight-learning techniques. IEEE Trans. Fuzzy Syst. **19**(4), 729–744 (2011)
9. D. Tikk, P. Baranyi, T.D. Gedeon, L. Muresan, Generalization of the rule interpolation method resulting always in acceptable conclusion. Tatra Mt. Math. Publ. **21**, 73–91 (2001)

Chapter 6
Hierarchical Bidirectional Fuzzy Rule Interpolation and Rule Base Refinement

For many practical intelligent decision-making applications, the "curse of dimensionality" is a serious problem; that is, the number of rules increases exponentially along with the number of input variables to the fuzzy inference system [1]. An effective way to deal with this problem is to use hierarchical fuzzy systems [2]. Another problem that fuzzy rule-based system faces is on the other end of the spectrum: there may not be sufficient historical data to support the creation of the needed rules that would cover the entire problem space, but only a "sparse rule base". Fortunately, fuzzy rule interpolation can be employed for dealing with this group of problems. A potentially more challenging case is the combination of both, i.e. there may be many input features to consider while only a sparse rule base to support inference [3]. However, for such a combined problem, situations may become even more complicated where certain crucial antecedents are absent from given observations. This is because missing antecedents may well be involved in the subsequent inference processes, causing the final conclusion not deducible. To address the underlying problem of performing fuzzy rule interpolation (FRI) for certain antecedent variables, an original technique for backward fuzzy rule interpolation (B-FRI) has been proposed [4], significantly extending the existing FRI techniques.

B-FRI can be employed to allow for rule interpolation to be carried out when certain antecedents of observation variables are absent, whereas conventional methods do not work. The missing antecedents may be inferred or interpolated using the known antecedents and given conclusion during the interpolative reasoning process. In this chapter, the theoretical work of hierarchical bidirectional fuzzy rule interpolation (HBFRI) is proposed to meet the aforementioned challenges, based on previous research [4–6]. It supports both interpolation and extrapolation that involve multiple intertwined fuzzy rules, with each possibly having multiple antecedent variables. HBFRI is herein implemented using the scale and move transformation-based fuzzy interpolative reasoning (T-FRI) [7, 8], owing to its popularity and availability (although other FRI methods may be adapted to serve as the alternative if preferred).

S. Jin et al., *Backward Fuzzy Rule Interpolation*,
https://doi.org/10.1007/978-981-13-1654-8_6

Importantly, due to the high incompleteness of knowledge bases and the imprecision of observations, rules used for fuzzy interpolation may become inconsistent. Such inconsistencies may result from incorrect observations, incorrect rules in the given rule base or defective interpolation procedures [9]. This problem may deteriorate while dealing with real-world problems, where the rules are typically irregular in nature (i.e. they may not always address the same antecedent variables). Therefore, HBFRI is designed in such way to include a novel rule base refinement mechanism that helps remove the inconsistencies in the rule base, no matter whether the antecedents are known or not.

The rest of this chapter is organised as follows. Section 6.1 presents the scale and move transformation-based fuzzy rule interpolation and backward fuzzy interpolation, along with detailed descriptions of a method for generating hierarchical fuzzy rule bases. The algorithm of hierarchical fuzzy interpolation approach is given in Sect. 6.1.3. The algorithm for the refinement of a fuzzy rule base involving hierarchical fuzzy rule interpolation is given in Sect. 6.3. Two illustrative numerical examples are presented in Sects. 6.2 and 6.4, respectively, to demonstrate the effectiveness of the proposed approach. Section 6.5 concludes the chapter.

6.1 Hierarchical Bidirectional Fuzzy Rule Interpolation

6.1.1 Representation of Intermediate Variables

Without losing generality, the output variable of each layer within a certain *HFS* is represented by [10]:

$$y_{l,p} = \sum_{j_1 j_2 \ldots j_{P_{l-1,p}} i_1 i_2 \ldots i_{Q_{l,p}}} U_{l,p} V_{l,p} * y_{l,p}^{j_1 j_2 \ldots j_{P_{l-1,p}} i_1 i_2 \ldots i_{Q_{l,p}}} \tag{6.1}$$

where

$$U_{l,p} = \prod_{k=1}^{P_{l-1,p}} \mu_{l,p,k}^{jk}(y_{l-1,p,k}) \tag{6.2}$$

and

$$V_{l,p} = \prod_{k=1}^{Q_{l-1,p}} \upsilon_{l,p,k}^{jk}(x_{l,p,k}) \tag{6.3}$$

with $y_{l,p}$ representing the output of the pth fuzzy subsystem in the lth layer; $y_{l,p}^{j_1 j_2 \ldots j_{P_{l-1,p}} i_1 i_2 \ldots i_{Q_{l,p}}}$ being the THEN part of $j_1 j_2 \ldots j_{P_{l-1,p}} i_1 i_2 \ldots i_{Q_{l,p}}$th fuzzy rule; $P_{l-1,p}$ being the total number of outputs from the $(l-1)$th layer to $F_{l,p}$; $Q_{l-1,p}$ being the total number of original input variables to $F_{l,p}$; and $\mu^{jk}_{l,p,k}(y_{l-1,p,k})$ and $\upsilon^{jk}_{l,p,k}(x_{l,p,k})$ are fuzzy membership functions for $y_{l-1,p,k}$ and $x_{l,p,k}$, respectively.

6.1.2 *Learning Algorithm*

For a standard fuzzy system, the least square method (LSM) is usually used to gain an optimal modelling result. However, it is not easy to apply LSM when developing a hierarchical fuzzy system because in many cases the intermediate variables have no physical meaning. A possible solution is to use gradient-descent techniques [10, 11], such as the error backpropagation algorithm, which is a popular method to optimise the parameters in hierarchical fuzzy systems. The parameter updating of the lower levels is based on the errors propagated back from the upper fuzzy layer (which are ultimately based on the exploitation of the error back-propagated from the final output). The gradient-descent learning algorithm is given as follows.

First, let $e(k)$ be the error between the actual output $y(k)$ and the hierarchical system output $y^{'}(k)$ at time k:

$$e(k) = y^{'}(k) - y(k) \tag{6.4}$$

and the errors propagated back are defined as:

$$e_p(k) = e_q(k) \times \frac{\partial y_q(k)}{\partial y_p(k)} \tag{6.5}$$

where $e_p(k)$ denotes the error of subfuzzy system $p, p \geq 1$, which is propagated from its immediate adjacent upper sublayer fuzzy system $q, q = p - 1$, and $e_q(k)$ is defined in the same manner.

The parameters $y_p^{j_1 j_2 \ldots j_{P_{q,p}} i_1 i_2 \ldots i_{Q_{q,p}}}(k)$ in gradient-descent learning are computed by

$$\begin{aligned} &y_p^{j_1 j_2 \ldots j_{P_{q,p}} i_1 i_2 \ldots i_{Q_{q,p}}}(k+1) \\ &= y_p^{j_1 j_2 \ldots j_{P_{q,p}} i_1 i_2 \ldots i_{Q_{q,p}}}(k) - \eta \times U_q(k) V_q(k) \times e_p \end{aligned} \tag{6.6}$$

where η is the learning rate, $U_q(k) = \prod_{i=1}^{P_q} \mu^{j_i}_{q,i}(k)$, and $V_q(k) = \prod_{i=1}^{Q_q} \mu^{j_i}_{q,i}(x_p, i)$.

Based on the above, the algorithm for learning a hierarchical rule base can be summarised below:

1. Choose the membership functions for each input variable, including both the original, real input variables of the system being modelled and the intermediate variables. An even partition for each input variable is assigned on their corresponding definition domain. The definition domain for the original input variables can be directly gained from the training data set. However, the definition domain for the

intermediate variables may not have any actual meaning and hence, may not be associated with any explicit definition domain. Thus, the definition domain for the intermediate variables can be assumed (and normalised) as [0, 1].

2. Choose initial parameters $y_{l,p}^{j_1 j_2 \ldots j_{P_{l-1,p}} i_1 i_2 \ldots i_{Q_{l,p}}}$ for the pth subfuzzy system of the lth level randomly. These parameters will be adjusted in the following steps.
3. Update the parameters $y_{l,p}^{j_1 j_2 \ldots j_{P_{l-1,p}} i_1 i_2 \ldots i_{Q_{l,p}}}$ with respect to each learning iteration k, for each given input-output pair (x^r, y^r), where r denotes the index of the training data.
4. Go to Step 3 with $r = r + 1$ if $r < T$, where T is the total number of training data in the training set.
5. Compute the accumulated error $E = \frac{1}{2} \times \sum_{r=1}^{T} (\hat{y}^r - y^r)^2$ and check if E is less than a prespecified small value ϵ, or if k is larger than a prespecified maximal iteration number K: if so, end the training process; else, go to Step 3 with $k = k + 1$.

6.1.3 Algorithm of HBFRI

Further to the outline as shown previously in Fig. 7.1, the HBFRI algorithm can be summarised in principle, as follows.

1. Determine the distances between the observed values for each input variable and the antecedents of each rule in the subrule base of the lowest layer, and choose the closest rules to construct the transformation parameters required for computing the interpolation through the subsequent layers.
2. Calculate the subconsequence for each sublayer using the multiple multi-antecedent rules interpolation method. If this subconsequence is not the output of the final layer, then the subconsequence will form the fuzzy term for the relevant input variable of the next layer.
3. Compute the output at the layer above from the terms of the intermediate variables, and the values of the original input variables of course, by iterating the first and second steps.

Thus, the hierarchical interpolative approach can be detailed below:

1. *Determine Closest Rules with Respect to Input Variables* When certain variables are present to an HFIU as input, the first task of the HFIU is to determine the closest rules for them. In general, the variables here not only refer to the original input variables involved in the observation, but may also include the intermediate variables which have been introduced by previous applications of HBFRI. Particularly, the distance $d_j^*, j = 1, \ldots, k$ between the fuzzy set A_j^i and the set A_k^* is calculated using Eq. 2.73. The n ($n \geq 2$) rules which have the minimum distance measures are chosen as the closest rules from the observation.

2. *Construct the Intermediate Rule* The computation and representation of the intermediate variable y_i, $(i = 1, 2, ..., K - 2)$ are the most important issue in any interpolation, hierarchical or not; there is no exception in HBFRI. The process of generating the intermediate variables of the HBFRIis, however, the same as outlined in Sect. 3.2.2, depending on whether forward or backward T-FRI is required.
3. *Carry out Scale Transformation* For each trapezoidal antecedent fuzzy value appearing in any of the N chosen rules, this step calculates two scale rates $\overline{s}^*_{A_j}$ and $\underline{s}^*_{A_j}$ according to Eq. 3.10, which rescale the top and bottom supports of $A_j^{'}$ with respect to the observation (or previously interpolated outcome) A_j^*, resulting in $A_j^{''}$. The corresponding bottom scale rate $\underline{s}_{B_j}$ and the intermediate top scale rate $\overline{s}_{B_j}{}'$ of the intermediate conclusion $B_j^{'}$ are obtained by averaging those computed for the antecedents. The final $\overline{s}_{B_j}$ is obtained from applying Eq. 2.81.
4. *Carry out Move Transformation* Using the move rate m_{A_j} as given in Eq. 2.86, $A_j^{''}$ is moved so that the final transformed fuzzy set matches the exact shape of the observed (or interpolated) value A_j^*. From this, m_{B_j} for the conclusion can be calculated according to Eq. 2.86. The final interpolated result B_j^* can now be calculated by applying the scale and move transformation to $B_j^{'}$, using the parameters $\underline{s}_{B_j}$, $\overline{s}_{B_j}$ and m_{B_j}.

6.2 Function Approximation and Experimental Evaluation

6.2.1 Experimental Setup

In this section, an illustrative example is used to demonstrate the proposed HBFRI approach. In particular, a function approximation problem with three input variables is considered:

$$y = f(x_1, x_2, x_3) = (1 + x_1{}^{0.5} + x_2{}^{-1} + x_3{}^{-1.5})^2$$

where six fuzzy sets are evenly defined for each input variable. The hierarchical structure is shown in Fig. 6.1, where HFIU1 and HFIU2 are the two subsystem units, i.e. the first layer and second layer of this HBFRI approximation model, respectively. For simulation, 341 samples are uniformly created on the three-dimensional problem space, 216 of which are used for training and the remaining 125 for testing.

According to Sect. 6.1.2, suppose that the learning rate is set as 0.0004, each original sublayer fuzzy system will contain 32 rules, with the system having a total of 72 rules. In each hierarchical fuzzy interpolation unit, to demonstrate the effect of running only a sparse rule subset, it is assumed that only four fuzzy rules are chosen to construct its original subrule base. These four rules jointly describe the

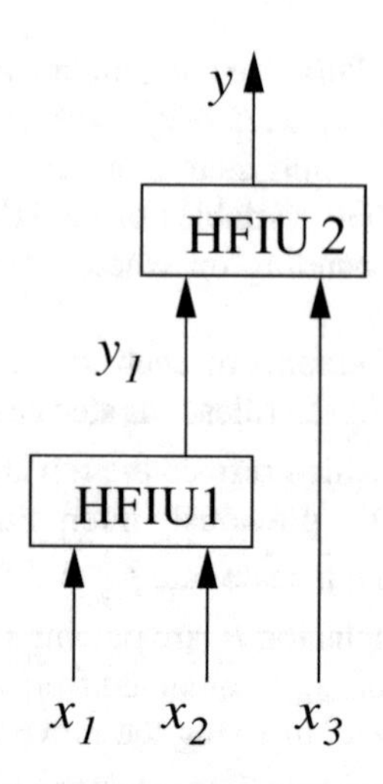

Fig. 6.1 Structure of illustrative hierarchical function approximation example

Table 6.1 Fuzzy subsparse rule base of the lower layer

Rule	Antecedents	Consequence
Rule 1	$x_1 = A_1^1 = (0.02, 0.51, 1.52, 2.01)$,	$y_1 = B_1^1 = (-0.24, 0.26, 1.26, 1.76)$
	$x_2 = A_2^1 = (0.03, 0.53, 1.53, 2.03)$	
Rule 2	$x_1 = A_1^2 = (0.02, 0.51, 1.51, 2.01)$,	$y_1 = B_1^2 = (0.00, 0.50, 1.50, 2.00)$
	$x_2 = A_2^2 = (4.99, 5.49, 6.49, 6.99)$	
Rule 3	$x_1 = A_1^3 = (4.99, 5.49, 6.49, 6.99)$,	$y_1 = B_1^3 = (-1.00, -0.50, 0.50, 1.00)$
	$x_2 = A_2^3 = (0.03, 0.53, 1.53, 2.03)$	
Rule 4	$x_1 = A_1^4 = (4.99, 5.49, 6.49, 6.99)$,	$y_1 = B_1^4 = (-0.65, -0.15, 0.85, 1.34)$
	$x_2 = A_2^4 = (4.99, 5.49, 6.49, 6.99)$	

Table 6.2 Fuzzy subsparse rule base of the upper layer

Rule	Antecedents	Consequence
Rule 1	$y_1 = B_1^1 = (-1.00, -0.50, 0.50, 1.0)$,	$y = B^1 = (20.76, 21.26, 22.26, 22.76)$
	$x_3 = A_3^1 = (0.01, 0.50, 1.503, 2.55)$	
Rule 2	$y_1 = B_1^2 = (-1.00, -0.50, 0.50, 1.00)$,	$y = B^2 = (14.90, 15.40, 16.40, 16.90)$
	$x_3 = A_3^2 = (4.98, 5.48, 6.48, 6.98)$	
Rule 3	$y_1 = B_1^3 = (0.00, 0.50, 1.50, 2.00)$,	$y = B^3 = (6.37, 6.87, 7.87, 8.37)$
	$x_3 = A_3^3 = (0.00, 0.50, 1.50, 2.00)$	
Rule 4	$y_1 = B_1^4 = (0.00, 0.50, 1.50, 2.00)$,	$y = B^4 = (3.20, 3.70, 4.70, 5.20)$
	$x_3 = A_3^4 = (4.98, 5.48, 6.48, 6.98)$	

minimum and maximum of the output and also those output values in response to the four points bounding the corresponding input space. Tables 6.1 and 6.2 display the subfuzzy sparse rules of the lower layer and those of the upper layer, respectively.

The approximation accuracy is measured by the average percentage error ($APE\%$) given as follows:

$$APE\% = (100/N)\sqrt{\sum_{i=1}^{N}(\frac{y_i^* - y_i'}{y_i'})} \tag{6.7}$$

where y_i^* is the ith objective output of the underlying function and y_i' is the ith model output.

6.2.2 Analysis of Results

Table 6.3 presents a comparative summary of running different models through different sets of data, including the standard fuzzy systems that use all the rules given in a flat set; the hierarchical fuzzy systems that use all a subset of rules arranged hierarchically; and the hierarchical interpolation system that only use a small portion of the rules employed in the hierarchical fuzzy systems.

Note that when six fuzzy sets are defined for each input variable, the maximum number (N_{max}) of involved fuzzy rules is $6^3 = 216$ for the standard fuzzy system, and $6^2 + 6^2 = 72$ for hierarchical fuzzy system, whereas the minimum number (N_{min}) of rules is $2 + 2 = 4$ for the proposed HBFRI. In order to attain the accuracy of inference while using fewer number of rules, in this experimentation, the cardinality of each sublayer rule base is taken to 4, so the total number of rules of the HBFRI is 8. To enhance the comparison, different set-ups where fewer rules are involved due to looser partitions of the three input variables are also tested, as reflected in Table 6.3.

Comparing the approximation results given in Table 6.3, the following observations can be made. Firstly, the hierarchical fuzzy interpolative approximation performs relatively well in the sense that it can achieve good approximation with significantly fewer rules. The accuracy of HBFRI is indeed between that of the standard

Table 6.3 Result comparison between standard and hierarchical fuzzy system

Method	Partitions	Rules	APE% testing
Standard fuzzy system	(5,5,5)	216	6.50119
Standard fuzzy system	(4,4,4)	125	6.37989
Hierarchical fuzzy system	(5,5,5)	72	1.48293
Hierarchical fuzzy system	(4,4,4)	32	1.71127
Hierarchical interpolation system	(5,5,5)	8	4.84735
Hierarchical interpolation system	(4,4,4)	8	5.76442

fuzzy system and that of the hierarchical fuzzy system, but the other two utilise substantially more fuzzy rules and hence require much more computational effort. Secondly, it can also be observed from the results on the APE% of the proposed hierarchical interpolation system that the more fuzzy terms are included to represent the system variables, and the more accurate the inference results may be obtained without increasing the number of rules used, showing the robustness of the approach. This is not the case for the other two types of fuzzy model (though the classical hierarchical technique also leads to more accurate results if more detailed descriptions on variables are used, but this requires a much large number of rules).

In addition, due to the underlying principle taken by the proposed approach for hierarchical bidirectional fuzzy rule interpolation, the sublayer rule bases can be constructed with less constraints over what intermediate variables to use such that the overall hierarchical model represents the original input–output problem space. This property enables HBFRI to allow for different settings for the original input variables to flow into the sublayers in any arbitrary order without affecting the final outcome. This may have been implicitly reflected in the results gathered in Table 6.3, but an explicit verification of this remains active research.

6.3 Refinement of Fuzzy Rule Base with HBFRI

Inference engine and fuzzy rule base are two key components of a fuzzy system, be it a conventional fuzzy system or a fuzzy interpolative system. Reasoning consistency may be assumed to exist in the original rule base. However, a new rule which is inferred or interpolated from the original rule base may lead to contradictory to an existing rule. In this chapter, HBFRI is employed to refine fuzzy rule bases when rule inconsistency appears.

Suppose that there are two inconsistent rules, respectively, represented as follows:

$$R_i : \text{IF } x_1 \text{ is } A_1^i, \cdots, x_k \text{ is } A_k^i, \cdots, x_M \text{ is } A_M^i, \\ \text{THEN } y \text{ is } B^i$$

$$R_i' : \text{IF } x_1 \text{ is } A_1^i, \cdots, x_k \text{ is } A_k^i, \cdots, x_M \text{ is } A_M^i, \\ \text{THEN } y \text{ is } B^{i'}$$

where $x_k, k = 1, 2, \cdots, M$, are the input variables, and $B^i \neq B^{i'}$. The process of refining is to re-compute a desired consequent value to replace both consequents in these rules. This can be implemented using the following procedure:

1. Backward interpolate each antecedent value A_k or $A_k^{'}$, respectively, from R_i or $R_i{'}$ using *B-FRI*, respectively.
2. Calculate the bias between each antecedent value A_k or $A_k^{'}$ and the corresponding observed antecedent value A_k^*.

$$\delta_k = \frac{|Rep(A_k) - Rep(A_k^*)|}{max_{A_k} - min_{A_k}}$$
$$\delta_k{'} = \frac{|Rep(A_k^{'}) - Rep(A_k^*)|}{max_{A_k} - min_{A_k}} \tag{6.8}$$

where max_{A_k} and min_{A_k} are the maximal and minimal domain values of the variable x_k, and $Rep(A)$ stands for the representative value of the fuzzy set A. The representative value for a trapezoidal fuzzy set $A = (a_0, a_1, a_2, a_3)$ has been defined in Eq. 2.56, where a_1 and a_2 denote the two bounding points of the nuclei of A, whose membership values are 1, and a_0 and a_3 denote the two extreme points of the support of A, whose membership values are 0.
3. Calculate the average bias for these two inconsistent rules, respectively.

$$\overline{\delta} = \frac{1}{M}\sum_{k=1}^{M} \delta_k$$
$$\overline{\delta'} = \frac{1}{M}\sum_{k=1}^{M} \delta_k{'} \tag{6.9}$$

4. The refined consequence is calculated from the above biases, resulting in the desired consequent to construct the new rule that replaces the original two inconsistent ones, with the same antecedent part as its originals, of course:

$$y = Rep(B^i) \times \frac{\overline{\delta'}}{\overline{\delta} + \overline{\delta'}} + Rep(B^{i'}) \times \frac{\overline{\delta}}{\overline{\delta} + \overline{\delta'}} \tag{6.10}$$

Note that the above description does not involve the execution of HBFRI, but the procedure for inconsistency removal. However, the discovery of the closest rules in the underlying implementation which leads to the production of interpolated rules assumes the use of the previously presented HBFRI scheme that represents rules with weighted consequent, reflecting the biased consideration towards the consequent variable.

6.4 Numerical Example-Based Evaluation

In this section, a simple example is used to illustrate the proposed approach. Here, a function approximation problem with three input variables is considered:

$$y = f(x_1, x_2, x_3) = (1 + {x_1}^{0.5} + {x_2}^{-1} + {x_3}^{-1.5})^2$$

where six fuzzy sets are defined for each input variable. For simulation, rules are uniformly created on the definition domain $U = [1, 6]^3$. The procedures of building the rule base and defining the fuzzy sets are omitted here to save space. Note that to demonstrate the success of sparse rule-based interpolative reasoning, only part of the constructed rule base is directly employed in this example, while including those rules that are created using previously interpolated results [12].

Suppose that there is a subrule base as shown in Table 6.4. For illustrative simplicity, all fuzzy values have been re-represented using their corresponding representative values. Without causing confusion, in the following discussion, when a value is mentioned it generally means the representative value of a fuzzy set unless otherwise stated.

In this table, $Rule_6$ and $Rule_{11}$ have the same antecedent values but different consequent values. This inconsistency of rules has to be eliminated. Firstly, the antecedent values $Rep(A_k)$ and $Rep(A_k^{'})$ $(k = 1, 2, 3)$ from R_6 and R_{11} can be obtained using Step 1) of the refinement procedure. Then, the bias between each antecedent value in these two rules and the averaged biases $\overline{\delta}$ and $\overline{\delta'}$ can be calculated according to Eqs. 6.8 and 6.9, with results as shown in Tables 6.5 and 6.6, respectively.

Table 6.4 Inconsistency example in a rule base

No. of rules	x_1	x_2	x_3	y
Rule 1	5.767954	5.735174	5.989867	13.280437
Rule 2	5.392889	1.326688	3.781224	17.741112
Rule 3	4.162835	3.670838	5.890916	11.442389
Rule 4	4.528533	2.403419	1.021757	20.361200
Rule 5	2.535111	5.767172	3.530740	8.504975
Rule 6	*5.656367*	*5.368069*	*5.674704*	*13.8392177*
Rule 7	3.895523	3.448394	4.953574	11.252007
Rule 8	1.752000	4.940259	5.221346	6.811400
Rule 9	2.814774	5.697204	5.370468	8.606024
Rule 10	3.348117	5.581531	1.977201	11.347698
Rule 11	*5.656367*	*5.368069*	*5.674704*	*10.565988*

Table 6.5 Deviation value of each antecedent for $Rule_6$

δ_1	δ_2	δ_3	$\overline{\delta}$
0.112183	0.060401645	0.036124967	0.06956986

Table 6.6 Deviation value of each antecedent for $Rule_{11}$

δ'_1	δ'_2	δ'_3	$\overline{\delta'}$
0.52580476	0.28467909	0.23902021	0.34983468

Table 6.7 The refined rule base

No. of rules	x_1	x_2	x_3	y
Rule 1	5.767954	5.735174	5.989867	13.280437
Rule 2	5.392889	1.326688	3.781224	17.741112
Rule 3	4.162835	3.670838	5.890916	11.442389
Rule 4	4.528533	2.403419	1.021757	20.361200
Rule 5	2.535111	5.767172	3.530740	8.504975
Rule 6	*5.656367*	*5.368069*	*5.674704*	*13.296262*
Rule 7	3.895523	3.448394	4.953574	11.252007
Rule 8	1.752000	4.940259	5.221346	6.811400
Rule 9	2.814774	5.697204	5.370468	8.606024
Rule 10	3.348117	5.581531	1.977201	11.347698

With respect to Eq. 6.10, the representative value of the final output of the refined rule is 13.296262, which is very close to the underlying ground truth value of 13.2392175. This accurate result demonstrates the significant potential of the proposed method in refining the interpolated rule base. Afterwards, $Rule_6$ and $Rule_{11}$ in the previous rule base are replaced by a new rule, resulting in Table 6.7.

6.5 Summary

This chapter has presented an initial approach for hierarchical bidirectional fuzzy rule interpolation and its use in support of decision-making. In particular, HBFRI enables unknown antecedent values to be interpolated, given other antecedents and the consequent. This integrated approach, of hierarchical reasoning and bidirectional interpolative inference, provides a flexible and systematic way of dealing with insufficient information or knowledge that may often appear in real-world problems. Furthermore, this chapter has proposed a new rule base refinement method, demonstrating that HBFRIcan also help resolve inconsistency potentially existing in the rule base.

To reveal the full potential of this work, an investigation into exactly how flexible in designing the hierarchical structure by introducing different intermediate variables requires further experimental research. Also, the present rule base-learning method may be expedited, while strengthening model interpretability if more advanced fuzzy rule learning techniques (e.g. [13]) may be integrated into the proposed algorithm. In addition, in this work, all rules are presented with every antecedent variable being weighted with equal significance, extending it to allowing for reasoning with weighting information would benefit from the selection of least number of the closest rules for interpolation [14], thereby further improving the algorithm efficiency. Last but not least, at the present, all fuzzy sets are prespecified to be in trapezoidal form. Using an automated data clustering tool such as those introduced in [15] would ensure more accurate interpolative results.

References

1. G. Raju, J. Zhou, A. Roger, Hierarchical fuzzy control. Int. J. Control **54**(5), 1201–1216 (1991)
2. G. Raju, J. Zhou, Adaptive hierarchical fuzzy controller. IEEE Trans. Syst. Man Cybern. **23**(4), 973–980 (1993)
3. L.T. Kóczy, K. Hirota, L. Muresan, Interpolation in hierarchical fuzzy rule bases, in *Proceedings of International Conference on Fuzzy Systems* (2000), pp. 471–477
4. L.T. Kóczy, K. Hirota, L. Muresan, Backward fuzzy rule interpolation. IEEE Trans. Fuzzy Syst. **22**(6), 1682–1698 (2014)
5. S. Jin, R. Diao, C. Quek, Q. Shen, Backward fuzzy rule interpolation with multiple missing values, in *Proceedings of IEEE International Conference on Fuzzy Systems* (2013), pp. 1–8
6. S. Jin, R. Diao, C. Quek, Q. Shen, Backward fuzzy interpolation and extrapolation with multiple multi-antecedent rules, in *Proceedings of IEEE International Conference on Fuzzy Systems* (2012), pp. 1170–1177
7. Z. Huang, Q. Shen, Fuzzy interpolative reasoning via scale and move transformations. IEEE Trans. Fuzzy Syst. **14**(2), 340–359 (2006)
8. Z. Huang, Q. Shen, Fuzzy interpolation and extrapolation: a practical approach. IEEE Trans. Fuzzy Syst. **16**(1), 13–28 (2008)
9. Z. Huang, Q. Shen, Adaptive fuzzy interpolation. IEEE Trans. Fuzzy Syst. **19**(6), 1107–1126 (2011)
10. D. Wang, X. Zeng, J. Keane, Intermediate variable normalization for gradient descent learning for hierarchical fuzzy system. IEEE Trans. Fuzzy Syst. **17**(2), 468–476 (2009)
11. D. Wang, X. Zeng, J. Keane, Analysis and design of hierarchical fuzzy systems. IEEE Trans. Fuzzy Syst. **7**(5), 617–624 (1999)
12. N. Naik, R. Diao, Q. Shen, Dynamic fuzzy rule interpolation and its application to intrusion detection. IEEE Trans. Fuzzy Syst. (2017)
13. T. Chen, C. Shang, P. Su, Q. Shen, Induction of accurate and interpretable fuzzy rules from preliminary crisp representation. Knowl. Based Syst. **146**, 152–166 (2018)

14. F. Li, C. Shang, Y. Li, J. Yang, Q. Shen, Fuzzy rule-based interpolative reasoning supported by attribute ranking, IEEE Trans. Fuzzy Syst. (2018)
15. T. Boongoen, C. Shang, N. Iam-On, Q. Shen, Extending data reliability measure to a filter approach for soft subspace clustering. IEEE Trans. Syst. Man Cybern Part B Cybern **41**(6), 1705–1714 (2011)

Chapter 7
Application: Terrorism Risk Assessment Using BFRI

7.1 Introduction

Terrorism and particularly suicide terrorist campaigns have became a high priority for governments, the media, and the general public. It is imperative to have a comprehensive security risk management programme including effective risk assessment and appropriate decision support for such activities. Terrorism risk assessment (TRA) therefore plays a crucial role in national and international security. In order to predict terrorist behaviour from a given set of evidence (including hypothesised scenarios), it is often necessary for investigators to reconstruct the possible scenarios that may have taken place. The difficulty of such constructed assessments lies with the inherent complexity and uncertainty of the underlying problem domain, which may be too challenging to directly comprehend [1]. Fuzzy reasoning-based systems [2, 3] can be beneficial in dealing with the lack of knowledge or information, particularly when concepts such as the "likelihood" of terrorist attacks become expressions involving uncertain qualitative values.

In the literature, a number of fuzzy reasoning-based TRA systems have been developed in an effort to assist in the task of combating terrorism, including [2–8]. A fuzzy decision-making support system may generate plausible scenarios so that investigators can analyse them hypothetically as well as objectively [6, 7]. Fuzzy inference systems have also been used to classify events related to terrorism [3]. In addition, a fuzzy belief/plausibility measure has been proposed [2] to capture the uncertainty of evaluating the risks of intentional terrorist acts, and a fuzzy ontology construction methodology regarding terrorism related activity is proposed in [3]. Approaches that evaluate terrorist scenarios using approximate reasoning have also been implemented in terms of linguistic belief [4, 5].

The use of the aforementioned fuzzy reasoning-based TRA systems has revealed two major challenges: high-dimensionality [9–13], and sparse rule base problems [14]. As described in Sect. 1.2, suppose there are K input variables and M membership functions for each variable, then M^K rules are required in order to construct a system that fully covers the underlying domain. This is usually referred to as the

S. Jin et al., *Backward Fuzzy Rule Interpolation*,
https://doi.org/10.1007/978-981-13-1654-8_7

rule-explosion problem or the curse of dimensionality. The rule-explosion problem can be addressed in two ways. The first is to reduce the number of fuzzy partitions M, which usually results in sparse rule bases [15–17]. Besides, for terrorism problem, there may not be sufficient historical data to support the creation of the needed rules that would cover the entire problem space. Then, naturally fuzzy interpolation can be employed for this group of problems. The other way is to reduce the dimensionality K of the subrule bases by using meta-levels or hierarchical fuzzy rule bases. An extreme case is the combination of both, which could improve the computational complexity dramatically [18].

In hierarchical systems, situations may become even more complicated where certain crucial antecedents may be absent from given observations. This is because such missing antecedents may be involved in the subsequent (subsystem) inference process, causing the final conclusion not deducible. To address the underlying problem of performing interpolation for certain antecedent variables, an initial technique for backward fuzzy rule interpolation [19] has been proposed. This allows the observations which directly relate to the conclusion to be inferred or interpolated from the other known antecedents and the given conclusion.

In this chapter, the initial theoretical work of BFRI is extended to meet the challenges where hierarchical bidirectional fuzzy rule interpolation (HBFRI) is needed. HBFRI is used to deal with practical applications which have a hierarchical framework. In these kinds of systems, the rules are typically irregular in nature (i.e., they may not always address the same antecedents). In particular, rules may be arranged in an interconnected mesh, where observations and conclusions in different subsets of rules may overlap, and yet may not be directly related throughout the entire rule base. For such complex systems, any missing values in a given set of observations may lead to failure in unidirectional interpolation.

HBFRI enables unknown antecedent values to be interpolated, given other antecedents and the conclusion. This integrated approach provides a flexible and systematic way of dealing with dynamic and insufficient information that may appear in a terrorism risk assessment process. The system implementing the proposed technique is able to draw a final conclusion through the exploitation of BFRI even when it is presented with partial observations. This helps identify hidden variables that may be useful during any subsequent decision support processes. In so doing, warnings may be issued and/or preventive measures may be deployed earlier.

HBFRI is implemented using scale and move transformation-based fuzzy interpolative reasoning (T-FRI) [20, 21]. According to previous research work [19, 22, 23], bidirectional (forward/backward) interpolation based on T-FRI is summarised, and the flowchart of the proposed method is presented in Fig. 7.1. T-FRI offers a flexible means to handle both interpolation and extrapolation involving multiple, multi-antecedent fuzzy rules. It guarantees the uniqueness, normality, and convexity of the resulting fuzzy sets. T-FRI is also able to handle various fuzzy set representations, including polygonal, Gaussian, and bell-shaped fuzzy membership functions.

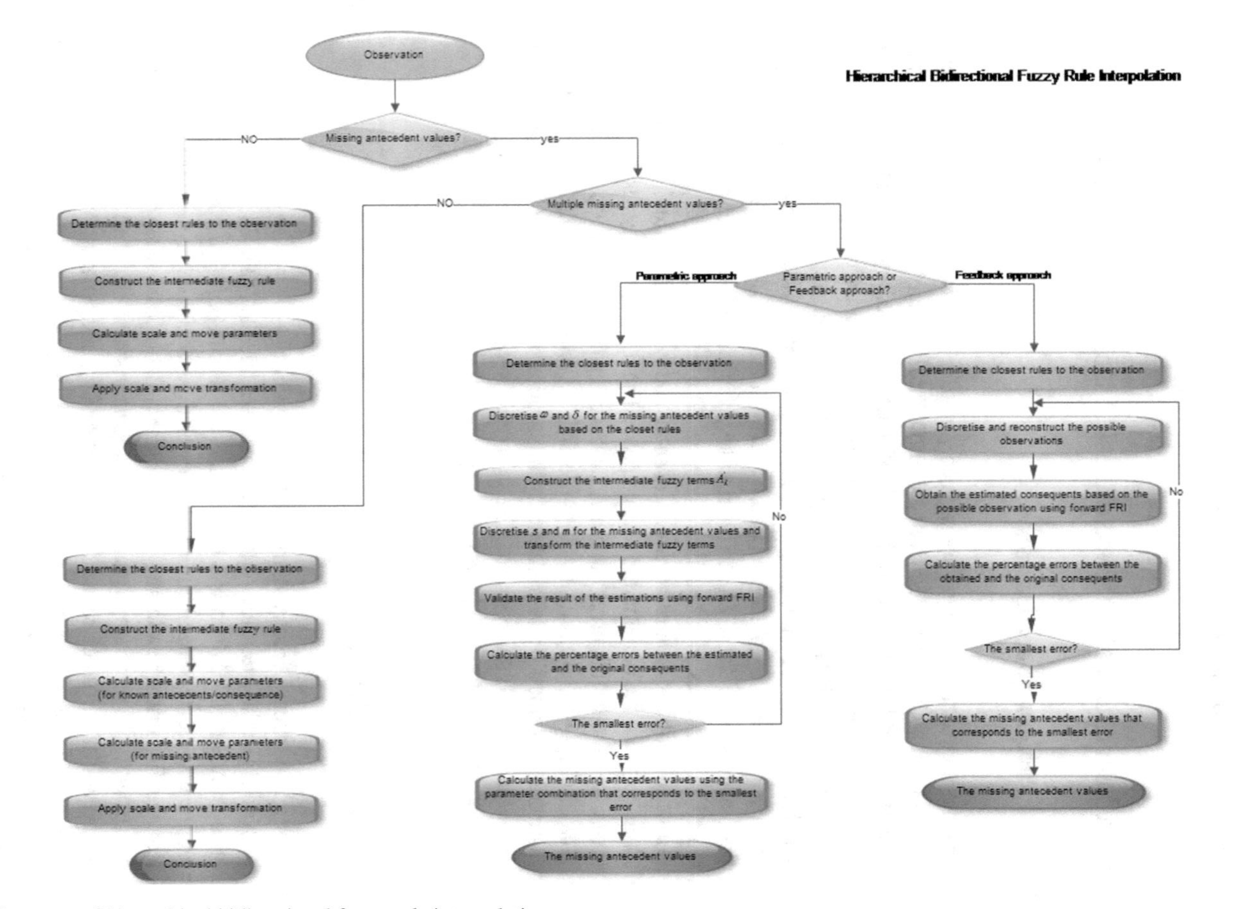

Fig. 7.1 Structure of hierarchical bidirectional fuzzy rule interpolation

The rest of this chapter is organised as follows. Section 7.2 presents the proposed framework for terrorism attack risk assessment. We provide a toy data set to illustrate the approach. In Sect. 7.3, the results of the experiments demonstrate the efficacy of the proposed approach. Section 7.4 concludes the chapter.

7.2 A Hierarchical Terrorism Risk Assessment Framework

7.2.1 *Problem Specification*

A terrorist attack never occurs by chance [24–26]. The framework of TRA proposed in this chapter is initiated from a comprehensive analysis of recent studies on counter-terrorism [24–32]. A bottleneck of TRA is caused by the fact that humans are relatively inefficient at hypothetical reasoning, especially when it comes to high-dimensional, hierarchically structured procedures involving uncertain data and sparse knowledge [33, 34]. In an attempt to address this problem, a four-layered analytical process for TRA is considered herein, as outlined in Fig. 7.2. The key variables involved are listed in Table 7.1. In the first layer, information about aspects directly attributed to terrorists is taken into account. The second layer is about the state government feedback, where *attrition*, *revenge*, and *negotiation* are the intermediate variables between the first and the second layer. The third layer regards the suspected terrorists, and the fourth layer concerns more localised information. This system is a typical multiple-input multiple-output (MIMO) system, which can be regarded as a construction formed by hierarchically integrating a collection of multiple-input single-output (MISO) subsystems. The following section provides a more concrete description of the individual layers of the proposed framework.

7.2.2 *Model Construction*

1. Layer 1: Aspects Directly Attributed to Terrorists
   According to the literature regarding terrorism analysis [24, 25, 29, 31, 32], the prime motivation of terrorism activities is for certain *political goals* (x_1), such as regime change, territorial change, policy change, social control, and status maintenance. Empirical results suggest that there is also a positive association between the growth rate of unemployment: *employment status* (x_2) and the incidence of terrorism [27]. Studies reveal that *education* (x_3) and *poverty* (x_4) in the regions have significant effects on terrorism also [28, 30]. The outputs of this layer are the levels of *attrition* (x_8), *negotiation* (x_{10}), and the *determination to terrorist activities* (x_{11}). As illustrated in Fig. 7.2, x_8 and x_{10} are utilised in layer two in order to provide decision-making support for state government, x_{11} is applied to decide the *state security level* (x_{19}). It also affects layer 4 (see later).

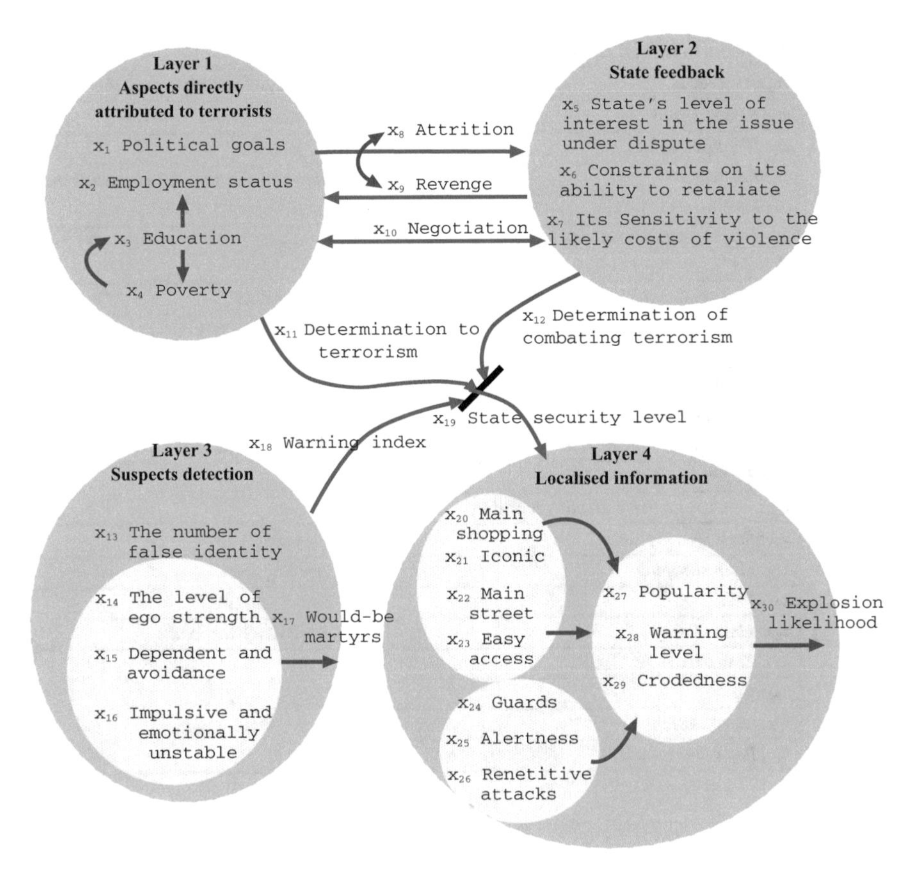

Fig. 7.2 The causal network model used

2. Layer 2: State Feedback
   This layer contains the responses to potential terrorist activities and corresponding reactions that may lead to such activities from the state government's viewpoint. The attitude and reactions taken by the state government mainly depend on the following factors [35]: *state's level of interest in the issue under dispute* (x_5), *constraints on its ability to retaliate* (x_6), *sensitivity to the likely costs of violence* (x_7), and the degree of *attrition* (x_8). These factors have significant impact upon the strength of *revenge* (x_9), or the government's position on the *negotiations* (x_{10}), and the government's *determination of combating certain terrorist activities* (x_{12}), which in turn is used to predict the *state security level* (x_{19}). Generally speaking, military retaliation is the most common response to terrorism attacks [35, 36], although occasionally *negotiation* with terrorists may be an alternative effective strategy [37–44], since a successful negotiation may save countless innocent lives [45, 46].

Table 7.1 Variables used in this TRA framework

	Input variables	Outputs and intermediate variables
Layer 1	x_1: *Political goals*	x_8: *Attrition*
	x_2: *Employment status*	x_{10}: *Negotiation*
	x_3: *Education*	x_{11}: *Determination to terrorist activities*
	x_4: *Poverty*	
Layer 2	x_5: *State's level of interest in the issue under dispute*	x_9: *Revenge*
	x_6: *Constraints on its ability to retaliate*	x_{10}: *Negotiation*
	x_7: *Its sensitivity to the likely costs of violence*	x_{12}: *Determination of combating terrorism*
	x_8: *Attrition*	
Layer 3	x_{13}: *The number of false identity*	x_{17}: *Would-be martyrs*
	x_{14}: *The level of ego strength*	x_{18}: *Warning index*
	x_{15}: *Dependent and avoidance*	
	x_{16}: *Impulsive and emotionally unstable*	
Layer 4	x_{19}: *State security level*	x_{27}: *Popularity*
	x_{20}: *Main shopping*	x_{28}: *Warning level*
	x_{21}: *Iconic*	x_{29}: *Crowdedness*
	x_{22}: *Main street*	x_{30}: *Explosion likelihood*
	x_{23}: *Easy access*	
	x_{24}: *Guards*	
	x_{25}: *Alertness*	
	x_{26}: *Repetitive attacks*	

Figure 7.3 illustrates an abstract example causal network for layers 1 and 2, involving 12 variables $x_i, i = 1, 2, ..., 12$. Six subrule bases are formed according to the background knowledge gained from the literature [24, 27–29, 42, 43]. For instance, one of the subsets of rules concerns with the concluding variable x_8 (which appears in this figure), explicitly involving logical conjunctions of 'antecedent' variables x_1, x_2, x_3, x_4, x_9 and x_{10}. This model helps provide decision support to the decision-makers regarding what kind of the responses (revenge or negotiation) may be adopted, and also the strength of revenge or the attitude of negotiation. Note that there exists a chain of responses between *attrition* and *negotiation* in chronological order.

3. Layer 3: Suspect Detection
This layer evaluates the mentalities of terrorists who might become suicide bombers or organisers of suicide attacks. Suicide attacks form the most salient tactic of present-day terrorism. They convey terrorists' willingness to kill indiscriminately and to 'sacrifice' themselves for their cause. Targets of such an attack may be certain iconic buildings, transportation interchanges, or the meeting places of important people. At border control, security services can detect suspects using various kinds of technology, e.g., false identity detection [47–49]. According to

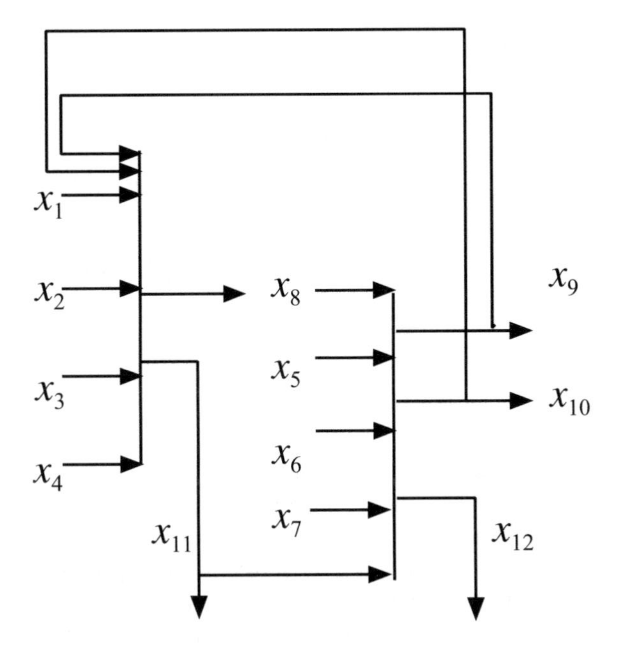

Fig. 7.3 Example causal network of the feedback for state government

the regular report of the *number of false identities* (x_{13}) given by the border control, further tests can be used to identify whether any suspects may be the real suicide bombers or organisers of suicide attacks. Personality test is also an effective means of detecting terrorist suspects [50, 51]. Other motivating factors may also induce or facilitate a person's readiness to be recruited for a 'martyrdom' mission. For example, *ego strength* (x_{14}) captures the person's ability to efficiently cope with external and internal stress, and to regulate one's emotions and needs.

In addition, there are (at least) two psychological factors: *dependent and avoidance* (x_{15}) and *impulsive and emotionally unstable* (x_{16}). People of a dependent and avoidance personality often long for interpersonal relationships from a position of low self-esteem. They are always in need of reassurance and may not function well without someone else to take care of them [51]. People with impulsive and emotionally unstable personality are characterised by emotional states that may change rapidly between melancholy, agitation, and hyperactivity. They show volatility in their occupational and social life [52]. In particular, recent results [51] show that members of 'martyrdom' groups are mostly characterised to be dependent and avoidance, while the majority of a commanding group have an impulsive and unstable personality. Two subrule bases regarding *would-be 'martyrs'* (x_{17}) and *warning index* (x_{18}) are formed according to studies in the literature [51, 53] and are illustrated in Fig. 7.4.

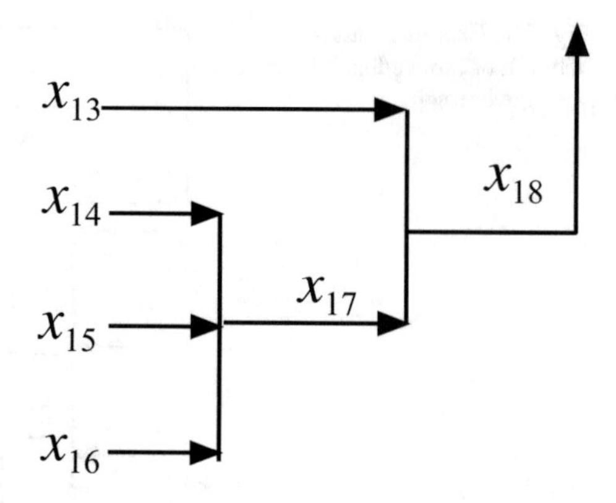

Fig. 7.4 Example causal network for detection of terrorism

4. Layer 4: Localised Information
   This layer focuses on the prediction of a terrorist attack and preventive control by adjusting certain regulatory factors in specific locations. To prevent attacks and protect lives, it is the most vital to implement a series of assessments on key locations. For instance, consider a practical scenario in detecting terrorist bomb threats. *Explosion likelihood* (x_{30}) may be directly related to the *warning level* (x_{28}) and the *crowdedness* (x_{29}) of a particular location. The number of people in an area directly determines the *popularity* (x_{27}) of the place and is itself affected by the level of the convenience of transportation *easy access* (x_{23}). *Main shopping* (x_{20}), *iconic* (x_{21}) and *main street* (x_{22}) are usually popular places for visitors. However, *crowdedness* (x_{29}) may change in relation to the *warning level* (x_{28}), since cautious individuals may shy away from areas that are considered to be dangerous.
   The number of public warning signs displayed in an area may discourage potential attacks, as people are more alert to their surroundings, and suspicious individuals or items may be promptly reported. Moreover, terrorists typically target certain places that repetitively draw their attention, rather than random locations [54]. Therefore, the *warning level* (x_{28}) of a local area, or the level of risk for the attackers, is related to the number of security *guards* (x_{24}) in the area, *alertness* (x_{25}) of people, and the numbers of *repetitive attacks* (x_{26}) in the past. It is also related to the *state security level* (x_{19}), which is decided by *determination to terrorist activities* (x_{11}), the government's *determination of combating terrorism* (x_{12}), and *warning index* (x_{18}).
   A causal network model of this layer is illustrated in Fig. 7.5, which contains 12 variables: $x_i, i = 19, \ldots, 30$, and four subrule bases.

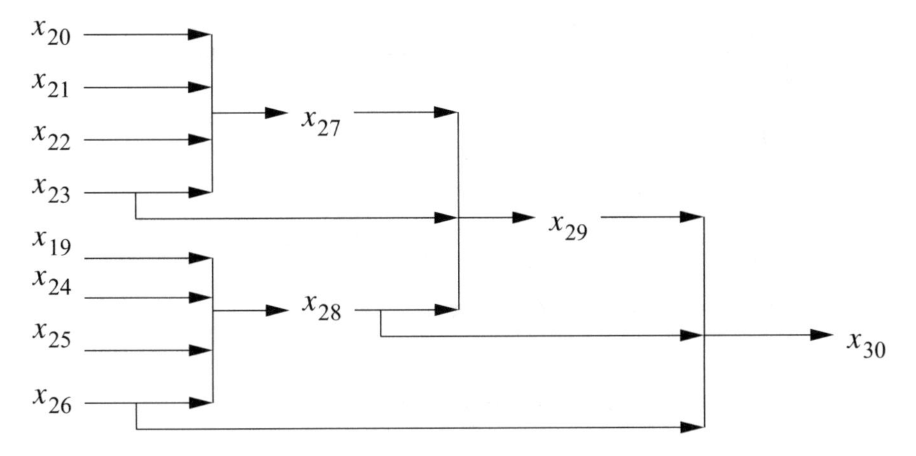

Fig. 7.5 Causal network model for the situation in a local government

7.3 Experimental Studies of TRA Using HBFRI

In this section, the TRA system described in Sect. 7.2 is implemented in order to demonstrate the potential of HBFRI. In such a real-world setting, it is often difficult to predict or adjust the strength of *revenge* (x_9), willingness to *negotiate* (x_{10}), *warning level* (x_{28}), and *guards* (x_{24}) unless terrorist attacks have actually occurred (or have been successfully prevented). However, with BFRI, these elements may now be estimated from other related factors. This may significantly increase the effectiveness of developing preventive means to counter terrorist attacks. As a result, security enhancements may be installed, particularly, for popular areas that pose an attractive target due to their strategic location.

7.3.1 *Simulated Assessment of Terrorist Risk Using HBFRI*

In the framework of Fig. 7.2, fuzziness is naturally obtained from the presence of the linguistic terms that describe the real-valued domain variables. For computational simplicity, triangular fuzzy membership functions are applied in this scenario. Note that different variables are defined upon their underlying domains. To simplify knowledge representation, these domains are normalised in this experiment to a range of 0–1. The fuzzy sets that represent the normalised linguistic terms are given in Fig. 7.6. Note also that the original rule base consists of substantial gaps which makes interpolation essential. A selection of the original rules contained in these subrule bases are shown in Tables 7.2, 7.3 and 7.4.

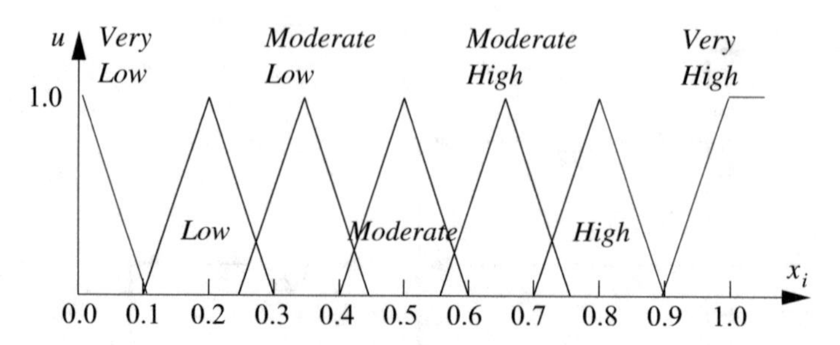

Fig. 7.6 Definition of the linguistic terms for domain variables

7.3.1.1 Prediction and Decision Support for State Government

Suppose that a set of observations O_1–O_6 are provided as given in Table 7.5, where the values of the antecedent variables *political goals* (x_1), *attrition* (x_8), *revenge* (x_9), *negotiation* (x_{10}) and *warning index* (x_{18}) are all missing. The task is to derive a value for the final consequent *state security level* (x_{19}), through interpolation. Using the hierarchical reasoning framework, the process of interpolating the final result *warring level of states* is summarised as follows.

(a) Calculate the value of *political goals* (x_1) according to the observed consequence value x_8 and the antecedent values x_2, x_3 and x_4, using the subrule base *Attrition* in Table 7.2 via BFRI. Following the steps listed in Fig. 7.1, the backward interpolated values can be computed such that $x_1 = (0.40, 0.50, 0.60)$ (*M*), $x_{10} = (0.64, 0.74, 0.84)$ (*MH*).

(b) Interpolate the value of x_{11}. Using the subrule base *Revenge* in Table 7.2 and observation O_2 in Table 7.5, the interpolated value of $x_9 = (0.65, 0.75, 0.85)$ can be obtained. The value of x_{11} is then inferred from x_9 and x_4, using the subrule base *Determination to terrorist activities* given in Table 7.2 and observation O_3 in Table 7.5. The resulting closest rules with respect to the observation are R_1 and R_3 which are listed in the subrule base *Determination to terrorist activities* in Table 7.2, and the interpolated value $x_{11} = (0.52, 0.62, 0.72)$ (*MH*).

(c) The other values of x_{10} and x_{12} are forward interpolated (through conventional T-FRI), using the subrule bases *Negotiation*, *Determination of combating terrorism* and observations O_4, O_5, respectively. This leads to the result $x_{10} = (0.58, 0.68, 0.78)$ (*MH*), $x_{12} = (0.41, 0.51, 0.61)$ (*M*). It can be seen that the value $x_{10} = (0.58, 0.68, 0.78)$ (*MH*) is consistent with the backward interpolated result value of $x_{10} = (0.64, 0.74, 0.84)$ (*MH*) obtained in step (a). The decision support given here is that the government's willingness for *negotiation* is not very strong, but the door for any negotiation would not be closed completely.

Table 7.2 Example rules for state's government

	Attrition					
	(x_1)	(x_2)	(x_3)	(x_4)	(x_{10})	(x_8)
R_1	VL	L	L	L	H	L
R_2	L	L	M	M	L	M
R_3	M	M	H	M	M	MH
R_4	H	H	H	H	VL	VH
	Revenge					
	(x_5)	(x_6)	(x_7)	(x_8)	(x_9)	
R_1	L	L	L	L	L	
R_2	L	M	M	M	M	
R_3	M	M	H	M	MH	
R_4	H	H	H	H	VH	
	Determination to terrorism					
	(x_1)	(x_2)	(x_3)	(x_4)	(x_9)	(x_{11})
R_1	L	H	L	L	L	ML
R_2	M	ML	M	M	M	MH
R_3	MH	M	MH	M	H	H
R_4	H	L	H	H	H	VH
	Negotiation					
	(x_5)	(x_6)	(x_7)	(x_8)	(x_{11})	(x_{10})
R_1	L	H	L	L	L	L
R_2	L	MH	M	M	M	M
R_3	M	M	H	M	MH	MH
R_4	H	L	H	H	H	VH
	Determination of combating terrorism					
	(x_5)	(x_6)	(x_7)	(x_8)	(x_{12})	
R_1	L	L	L	L	L	
R_2	ML	ML	M	M	M	
R_3	M	M	MH	M	MH	
R_4	H	H	VH	H	VH	
	State security level					
	(x_{11})	(x_{12})	(x_{18})	(x_{19})		
R_1	L	L	L	ML		
R_2	L	M	M	MH		
R_3	M	H	M	M		
R_4	H	H	H	VH		

VL: Very Low, L: Low, ML: Moderate Low, M: Moderate, MH: Moderate High, H: High, VH: Very High

Table 7.3 Example rules for terrorists detection

	Would-be martyrs			
	(x_{14})	(x_{15})	(x_{16})	(x_{17})
R_1	L	L	L	H
R_2	L	M	M	MH
R_3	M	H	M	M
R_4	H	H	H	L
	Warning index			
	(x_{13})	(x_{17})	(x_{18})	
R_1	L	L	L	
R_2	M	M	M	
R_3	H	M	MH	
R_4	H	H	VH	

VL: Very Low, L: Low, ML: Moderate Low, M: Moderate, MH: Moderate High, H: High, VH: Very High

7.3.1.2 Prediction and Decision Support for Suspects Detection

Assume that a set of observations O_7–O_8 are presented in Table 7.5, which is reportedly given by the border control. The value of the antecedent variable *number of false identities* (x_{13}) is $(0.4, 0.5, 0.6)$. The value of the variable *ego strength* (x_{14}) of a certain suspect is $(0.7, 0.8, 0.9)$. The values of the two psychological factors x_{15} and x_{16} are $(0.55, 0.65, 0.75)$ and $(0.7, 0.8, 0.9)$, respectively. In O_7, the value of the antecedent variable *would-be 'martyrs'* (x_{17}) is missing. Hence, the consequent x_{18} cannot to be directly interpolated. Using HBFRI, the results is as follows: $x_{17} = (0.80, 0.90, 0.95)$ (*VH*), which means this suspect is more likely be a real 'martyr' or organiser of suicide attacks. The results regarding prediction of alert level: *warning index* $(x_{18}) = (0.70, 0.80, 0.90)$ (*H*), and $x_{19} = (0.80, 0.90, 0.95)$ (*VH*) (calculated by using observation O_6 and the subrule base *State security level*). These show that the security alert level is very high for all aspects.

7.3.2 *Prediction and Decision Supporting for Local Government*

For local government, a set of observations O_9–O_{12} which is used in this experiment is given in Table 7.5, where the values of the antecedent variables *Easy access* (x_{23}), *Guards* (x_{24}), and *Warning level* (x_{28}) are all missing, as well as that of the final consequent x_{30} to be interpolated. The antecedent variable *Warning level* (x_{28}) is of particular importance, since it is involved in two subsets of rules (for *Crowdedness* and *Explosion likelihood*). Without x_{28}, no matter what other information is

Table 7.4 Example rules for local governments

	Popularity				
	(x_{20})	(x_{21})	(x_{22})	(x_{23})	(x_{27})
R_1	VL	L	L	L	L
R_2	L	H	L	M	ML
R_3	M	H	M	MH	M
R_4	VH	H	H	M	H
	Warning level				
	(x_{19})	(x_{24})	(x_{25})	(x_{26})	(x_{28})
R_1	L	L	VL	VL	VL
R_2	M	L	H	M	M
R_3	L	M	L	M	ML
R_4	H	H	H	H	VH
	Crowdedness				
	(x_{23})	(x_{27})	(x_{28})	(x_{29})	
R_1	VL	VL	VH	VL	
R_2	H	M	M	MH	
R_3	ML	H	L	M	
R_4	H	H	VL	VH	
	Explosion likelihood				
	(x_{26})	(x_{28})	(x_{29})	(x_{30})	
R_1	VL	L	L	VL	
R_2	L	M	M	L	
R_3	H	M	M	MH	
R_4	VH	M	L	H	

VL: Very Low, L: Low, ML: Moderate Low, M: Moderate, MH: Moderate High, H: High, VH: Very High

available, even if *Repetitive attack* (x_{26}) and *Crowdedness* (x_{29}) are known, forward interpolation will still fail.

In this hierarchical reasoning framework, the process of interpolating the final result *Explosion likelihood* (x_{30}) is carried out as follows. It is not uncommon for a hierarchical reasoning framework to have more than one path of inference/interpolation [55, 56]. For this particular set of observations, it is possible to obtain the value of x_{28} through two different paths, as shown in Fig. 7.7.

- *Dotted Path via S-BFRI*:

(a) Calculate the value of *Easy access* (x_{23}) according to the given consequence value x_{27} and the antecedent values x_{20}, x_{21} and x_{22}, using the subrule base *Popularity* via S-BFRI. Following the steps detailed previously, the backward interpolated value is $x_{23} = (0.42, 0.52, 0.62)$ (M).

Table 7.5 Observations

O_1	x_1	x_2	x_3	x_4	x_{10}	x_8
	missing	L	L	H	missing	MH
O_2	x_5	x_6	x_7	x_8	x_9	
	M	MH	MH	MH	?	
O_3	x_1	x_2	x_3	x_4	x_9	x_{11}
	L	L	H	missing	missing	?
O_4	x_5	x_6	x_7	x_8	x_{11}	x_{10}
	M	MH	MH	MH	missing	?
O_5	x_5	x_6	x_7	x_8	x_{12}	
	M	MH	MH	MH	?	
O_6	x_{11}	x_{12}	x_{18}	x_{19}		
	missing	missing	missing	?		
O_7	x_{14}	x_{15}	x_{16}	x_{17}		
	H	MH	H	?		
O_8	x_{13}	x_{17}	x_{18}			
	M	missing	?			
O_9	x_{20}	x_{21}	x_{22}	x_{23}	x_{27}	
	H	H	MH	missing	H	
O_{10}	x_{19}	x_{24}	x_{25}	x_{26}	x_{28}	
	missing	missing	L	MH	?	
O_{11}	x_{23}	x_{27}	x_{28}	x_{29}		
	missing	H	missing	M		
O_{12}	x_{26}	x_{28}	x_{29}	x_{30}		
	ML	missing	M	?		

(b) Interpolate the value of x_{28} using the subrule base *Crowdedness*. The values of x_{27} and x_{29} are directly observed, and that of x_{23} is obtained from the previous step. The three closest rules are then selected using the consequence-biased distance measure as per Eq. 3.5. The resulting closest rules with respect to the observation, and the backward interpolated value $x_9 = (0.52, 0.62, 0.72)$ (*MH*), are shown in Fig. 7.8.

(c) Use the interpolated value of x_{28}, and the other given values for x_{26} and x_{29} to forward interpolate the final consequent variable *Explosion likelihood* (x_{30}). This leads to the required final result $x_{30} = (0.61, 0.71, 0.81)$ (*MH*).

- *Dashed Path via P-BFRI/F-BFRI*

(a) Calculate x_{23} and x_{28} simultaneously, according to the given values x_{27} and x_{29}, and the subrule base *Crowdedness*. By employing the P-BFRI approach, the results obtained are as follows: $x_{23} = (0.46, 0.56, 0.66)$ (*M*) and $x_{28} = (0.51, 0.61, 0.71)$ (*MH*). Alternatively, the results obtained using F-BFRI are as

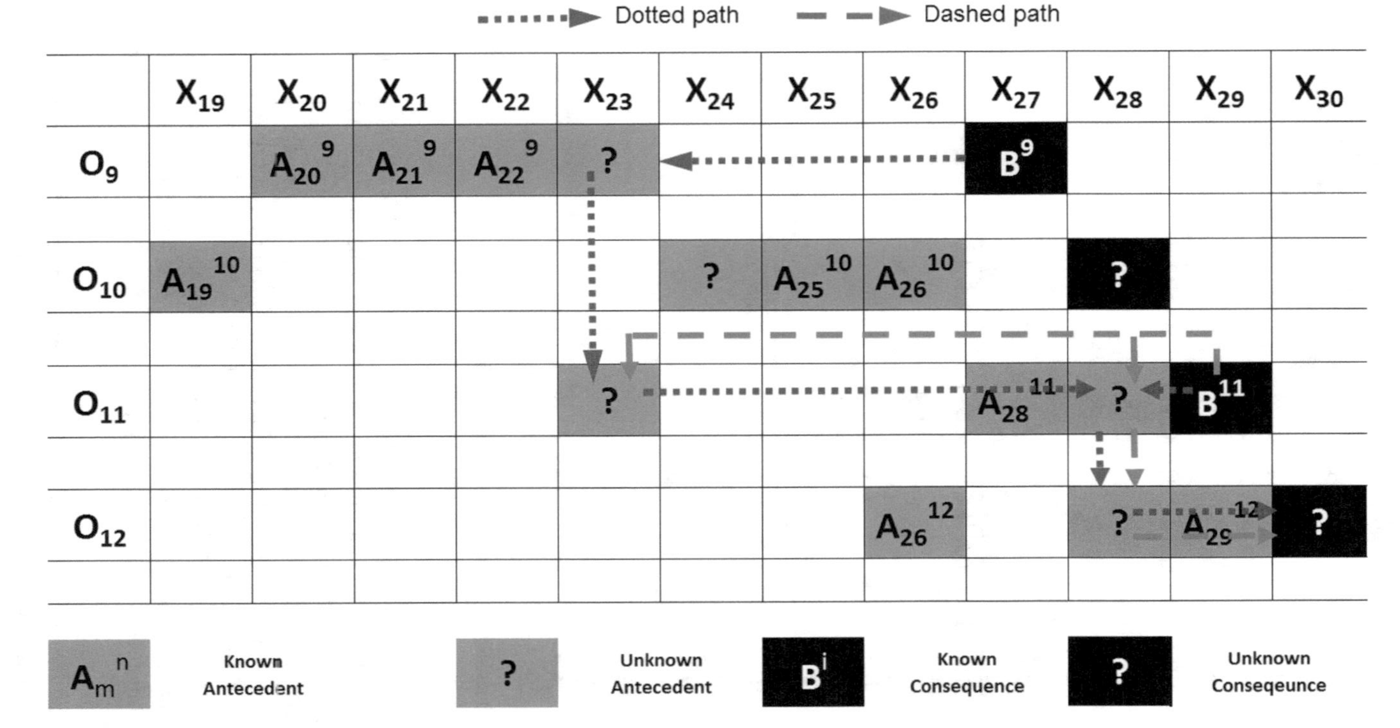

Fig. 7.7 Example structure for bombing attack prediction

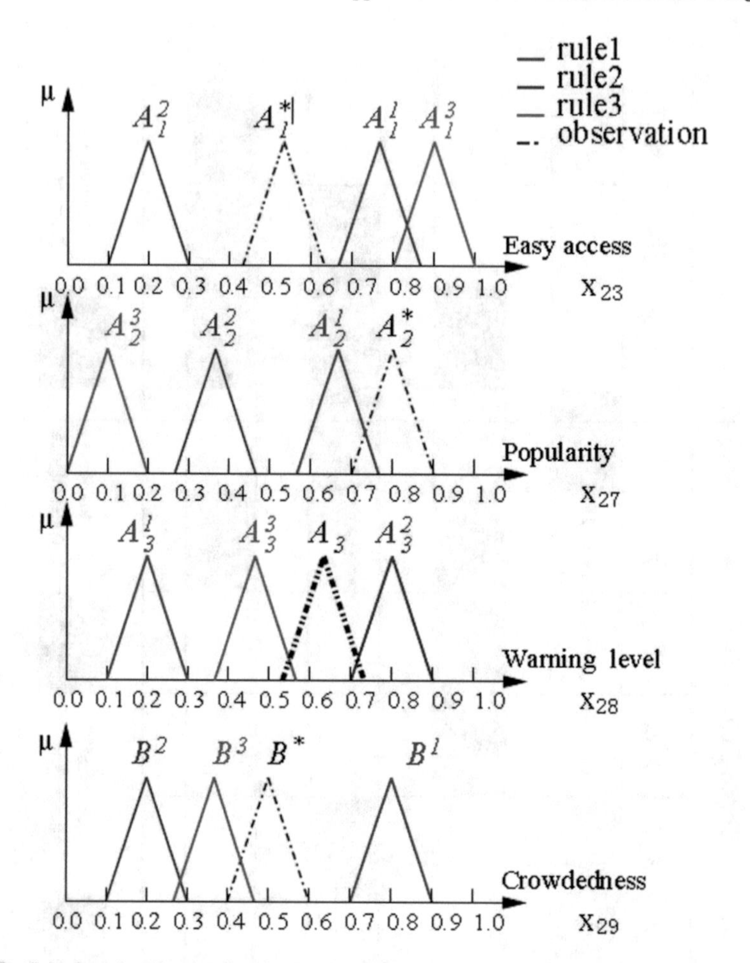

Fig. 7.8 Calculate the *Warning level* using S-BFRI

follows: $x_{23} = (0.43, 0.53, 0.63)$ (*M*) and $x_{28} = (0.59, 0.69, 0.79)$ (*MH*). Both agree with the previous results via the use of S-BFRI: $x_{23} = (0.42, 0.52, 0.62)$ (*M*), $x_{28} = (0.52, 0.62, 0.72)$ (*MH*).

(b) Derive, in a manner similar to the dotted path above, the value of the final consequent variable $x_{30} = (0.58, 0.68, 0.78)$ (*MH*) (according to P-BFRI: $x_{28} = (0.51, 0.61, 0.71)$) and $x_{30} = (0.56, 0.66, 0.76)$ (*MH*) (according to F-BFRI: $x_{28} = (0.59, 0.69, 0.79)$). The error $\epsilon_{\%}$ between the result of the dotted path ($x_{30} = (0.61, 0.71, 0.81)$) and these are 3.0% and 5.0%, respectively. This demonstrates that both processing paths are feasible for dealing with this problem.

- Note that after the aforementioned interpolation process, the last remaining missing value for *Guards* (x_{24}), although of no use in this particular prediction application, can also be backward interpolated. In particular, by using the observed values of x_{25} and x_{26}, and the interpolated value of x_{28}, the result of $x_{24} = (0.28, 0.38, 0.48)$ (*ML*) can be obtained via the use of subrule base *Warning level.*

7.3.3 Practical Significance of BFRI

In real applications, it is often difficult to predict and adjust the *Warning level* until (suicide) bomb attacks have actually occurred or been prevented. However, with BFRI, the *Warning level* may now be estimated from the other related factors. This may significantly increase the effectiveness of the prediction and prevention of bombing attacks. However, *Easy access* and *Guards* are controllable elements which may be adjusted in order to minimise the *Explosion likelihood.*

In order to reduce *Explosion likelihood* (x_{30}) from $(0.61, 0.71, 0.81)$ (*MH*) say, to $(0.30, 0.40, 0.50)$ (*ML*), according to the proposed P-BFRI method, the value of *Easy access* (x_{23}) needs to be changed from the current $(0.42, 0.52, 0.62)$ (*M*) to $(0.23, 0.33, 0.43)$ (*ML*), and similarly, *Warning level* (x_{28}) from the current $(0.52, 0.62, 0.72)$ (*MH*) to $(0.74, 0.83, 0.94)$ (*H*), and *Guards* (x_{24}) from the current $(0.28, 0.38, 0.48)$ (*ML*) to $(0.54, 0.64, 0.74)$ (*MH*). Thus, with the use of the proposed reverse reasoning technique to interpolate the crucial variables, the risk of a certain area concerned may be significantly reduced (or future repetitive attacks prevented).

7.3.4 Use of Alternative Distance Metrics

If the unbiased distance measure (Eq. 5.15) is used to backward interpolate the missing value x_{28}, the closest rules will no longer be the same. Instead a different outcome of *Warning level* $x_{28} = (0.18, 0.28, 0.38)$, and *Explosion likelihood* $x_{30} = (0.29, 0.39, 0.49)$ (*ML*) will be returned. Looking back at the original observation, given the two antecedent values such that *Easy access* (x_{23}) is *M* and *Popularity* (x_{27}) is *H*, the intuitive deduction of *Crowdedness* (x_{29}) should be quite high, as the place is both moderately high in popularity and reasonably convenient to reach. The only reason why observed *Crowdedness* (x_{29}) has a *moderate* value may well be a result of a reasonable *Warning level*. Therefore, the outcome $x_{28} = (0.52, 0.62, 0.72)$ from the use of the biased distance measure is more agreeable than $x_{28} = (0.18, 0.28, 0.38)$ resulting from the use of the plain distance measure. Experiments show that if $x_{28} = (0.18, 0.28, 0.38)$ and the antecedent values of x_{23} and x_{27} are *M* and *H*, respectively, T-FRI method will result in an interpolated $x_{29} = (0.14, 0.24, 0.34)$ (*L*), which will be much further than the original observation: *Crowdedness* (x_{29}) is *M*. This clearly demonstrates the significance in utilising the biased distance metric proposed in this work.

Table 7.6 Observation used for the investigation of multiple possible outcomes

Main shopping	Iconic	Main street	Easy access	Popularity
(x_{20})	(x_{21})	(x_{22})	(x_{23})	(x_{27})
missing	missing	M	ML	MH

Table 7.7 Relationship between *Main shopping* and *Iconic*

$\epsilon_{\%}$	Main shopping		Iconic	
0.29	(0.66,0.76,0.86)	H	(0.50,0.60,0.70)	MH
0.56	(0.54,0.54,0.64)	M	(0.72,0.82,0.92)	H
0.91	(0.19,0.29,0.39)	VL	(0.76,0.86,0.96)	H
1.06	(0.50,0.60,0.70)	MH	(0.74,0.84,0.94)	H
1.20	(0.67,0.77,0.87)	H	(0.70,0.80,0.90)	H
1.28	(0.70,0.80,0.90)	H	(0.20,0.30,0.40)	ML

7.3.5 *Multiple Equally Probable Interpolative Outcomes*

The involvement of multiple missing antecedents naturally implies that alternative, equally probable combinations of observations may be present, which may all lead to the same consequent observed. Assume that the observation shown in Table 7.6 is given, where the values of *Main shopping* and *Iconic* are both missing. The P-BFRI is adopted in order to calculate these two missing antecedent values, and the different results found with similar errors ($\epsilon_{\%} < 1.50\%$) are summarised in Table 7.7. These results reveal that the same value of *Popularity* can be obtained when either *Main shopping* or *Iconic* is *High* or *Very High*. This also agrees with the intuition, since either of these variables may lead to a given location to be seen as an attractive target, and both may be equally effective in influence the final outcome. Theoretically speaking, this particular subset of rules contains redundant variables and may be further pruned for higher efficiency. The presence of such redundancy may prompt the use of dimensionality reduction techniques such as feature selection [57, 58].

7.4 Summary

This chapter has addressed the applicability and usefulness of the proposed BFRI approach for terrorism risk assessment and decision-making support. Terrorist attacks launched by extremist groups or individuals have caused catastrophic consequences worldwide. Terrorism risk assessment therefore plays a crucial role in national and international security. Fuzzy reasoning-based terrorism risk assessment systems offer significant potential for providing decision-making support in combating terrorism, particularly for highly complex situations are involved. However, missing expertise often presents challenges for configuring systems that can otherwise assess the likelihood and risk of possible attacks due to the availability of only sparse rule bases. Hierarchical fuzzy rule interpolation systems may be adopted in order to overcome such problems. Unfortunately, situations can become more sophisticated because certain important antecedent values may be missing, which need to be inferred from the known (or hypothesised) consequences. Initial theoretical work to backward fuzzy rule interpolation has been proposed in order to address these underlying problems.

Systematic analysis of a hierarchical terrorism risk assessment framework has been given. A four-layered analytical process with a detailed description for TRA is considered. This chapter presents such an integrated approach that is capable of dealing with dynamic and insufficient information in the risk assessment process. In particular, the hierarchical system implementing the proposed technique can both predict the likelihood of terrorist attacks on different segments of focussed attention, and also help identify hidden variables that may be useful during the decision support process (by performing reverse inference). This chapter also reported on an experimental investigation of this implemented system, prediction and decision support for state government, suspects detection, and local government. Although not all of the rules are extracted from a credible data set, and it could be viewed as a toy system. However, the results demonstrate the potential and the efficacy of the HBFRI approach. The topics of practical significance of this implemented system, alternative distance metrics, and equally probable outcomes have also been discussed.

References

1. B.C. Ezell, S.P. Bennett, D. Von Winterfeldt, J. Sokolowski, A.J. Collins, Probabilistic risk analysis and terrorism risk. Risk Anal. **30**(4), 575–589 (2010)
2. J.L. Darby, *Evaluation of Risk from Acts of Terrorism: The Adversary/defender Model using Belief and Fuzzy Sets* (United States, Department of Energy, 2006)
3. U. Inyaem, P. Meesad, C. Haruechaiyasak, D. Tran, Terrorism event classification using fuzzy inference systems (IJCSIS). Int. J. Comput. Sci. Inf. Secur. **7**(3) (2010)
4. J.B. Bowles, C.Enrique Peláez, Fuzzy logic prioritization of failures in a system failure mode, effects and criticality analysis. Reliab. Eng. Syst. Saf. **50**(2), 203–213 (1995)
5. J.B. Bowles, Peláez, C. Enrique, *Linguistic Evaluation of Terrorist Scenarios: Example Application*, SAND2007-1301 (Sandia National Laboratories, Albuquerque, NM, 2007)
6. X. Fu, Q. Shen, Fuzzy compositional modeling. IEEE Trans Fuzzy Syst. **18**(4), 823–840 (2010)
7. Q. Shen, J. Keppens, C. Aitken, B. Schafer, M. Lee, A scenario-driven decision support system for serious crime investigation. Law Probab. Risk **5**(2), 87–117 (2006)
8. W.L. Waugh, *International Terrorism: How Nations Respond to Terrorists* (Documentary Publications Salisbury, NC, 1982)
9. R. Alcala, M.J. Gacto, F. Herrera, A fast and scalable multiobjective genetic fuzzy system for linguistic fuzzy modeling in high-dimensional regression problems. IEEE Trans. Fuzzy Syst. **19**(4), 666–681 (2011)
10. J. Mostafa, S. Mukhopadhyay, M. Palakal, W. Lam, A multilevel approach to intelligent information filtering: model, system, and evaluation. ACM Trans. Inf. Syst. (TOIS) **15**(4), 368–399 (1997)

11. U. Kaymak , R. Babuska, Compatible cluster merging for fuzzy modeling, in *Proceedings of the FUZZ-IEEE/IFES95* (1995), pp. 897–904
12. B. Song, R. Marks, S. Oh, P. Arabshahi, T. Caudell, J. Choi, et al., Adaptive membership function fusion and annihilation in fuzzy if-then rules, in *Second IEEE International Conference on Fuzzy Systems* (IEEE, 1993), pp. 961–967
13. C. Sun, Rule-base structure identification in an adaptive-network-based fuzzy inference system. IEEE Trans. Fuzzy Syst. **2**(1), 64–73 (1994)
14. B. John Garrick, J.E. Hall, M. Kilger, J.C. McDonald, T. O'Toole, P.S. Probst, E. Rindskopf Parker, R. Rosenthal, A.W. Trivelpiece, L.A. Van Arsdale, et al., Confronting the risks of terrorism: making the right decisions. Reliab. Eng. Syst. Saf. **86**(2), 129–176 (2004)
15. M. Sugeno, G. Kang, Fuzzy modeling and control of multilayer incinerator. Fuzzy Sets Syst. **18**, 329–346 (1986)
16. M. Sugeno, I. Hirano, S. Nakamura, S. Kotsu, Development of an intelligent unmanned helicopter, in *Proceedings of International Conference on Fuzzy Systems*, vol. 5 (IEEE, 1995), pp. 33–34
17. M. Sugeno, G. Kang, Structure identification of fuzzy model. Fuzzy Sets Syst. **28**(1), 15–33 (1988)
18. L.T. Kóczy, K. Hirota, L. Muresan, Interpolation in hierarchical fuzzy rule bases, in *Proceedings of International Conference on Fuzzy Systems* (2000), pp. 471–477
19. L.T. Kóczy, K. Hirota, L. Muresan, Backward fuzzy rule interpolation. IEEE Trans. Fuzzy Syst. **22**(6), 1682–1698 (2014)
20. Z. Huang, Q. Shen, Fuzzy interpolative reasoning via scale and move transformations. IEEE Trans. Fuzzy Syst. **14**(2), 340–359 (2006)
21. Z. Huang, Q. Shen, Fuzzy interpolation and extrapolation: a practical approach. IEEE Trans. Fuzzy Syst. **16**(1), 13–28 (2008)
22. S. Jin, R. Diao, C. Quek, Q. Shen, Backward fuzzy rule interpolation with multiple missing values, in *Proceedings of IEEE International Conference on Fuzzy Systems* (2013), pp. 1–8
23. S. Jin, R. Diao, C. Quek, Q. Shen, Backward fuzzy interpolation and extrapolation with multiple multi-antecedent rules, in *Proceedings of IEEE International Conference on Fuzzy Systems* (2012), pp. 1170–1177
24. S. Atran, Genesis of suicide terrorism. Science **299**(5612), 1534–1539 (2003)
25. R.A. Pape, The strategic logic of suicide terrorism. Am. Polit. Sci. Rev. **97**(3), 343–361 (2003)
26. W.L. Perry, C. Berrebi, R.A. Brown, J. Hollywood, A. Jaycocks, P. Roshan, T. Sullivan, L. Miyashiro, *Predicting Suicide Attacks: Integrating Spatial, Temporal and Social Features of Terrorist Attack Targets* (RAND Corporation, Santa Monica, CA, 2013)
27. R. Caruso, E. Gavrilova, Youth unemployment, terrorism and political violence, evidence from the israeli/palestinian conflict. Peace Econ. Peace Sci. Publ. Policy **18**(2) (2012)
28. E.B. De Mesquita, The quality of terror. Am. J. Polit. Sci. **49**(3), 515–530 (2005)
29. D. Fromkin, The strategy of terrorism. Foreign Aff. **53**(4), 683–698 (1975)
30. A.B. Krueger, J. Maleckova, Education, poverty and terrorism: is there a causal connection. J. Econ. Perspect. **17**(4), 119–144 (2003)
31. G. LaFree, L. Dugan, Introducing the global terrorism database. Terror. Polit. Violence **19**(2), 181–204 (2007)
32. M.A. Sutton, O. Oenema, J.W. Erisman, A. Leip, H. van Grinsven, W. Winiwarter, Too much of a good thing. Nature **472**(7342), 159–161 (2011)
33. B.P. Bryant, R.J. Lempert, Thinking inside the box: a participatory, computer-assisted approach to scenario discovery. Technol. Forecast. Soc. Change **77**(1), 34–49 (2010)
34. V.A. Epanechnikov, Non-parametric estimation of a multivariate probability density. Theory Probab. Appl. **14**(1), 153–158 (1969)
35. K. H, B.F. Walter, The strategies of terrorism. Int. Secur. **31**(1), 49–80 (2006)
36. L. Huddy, S. Feldman, C. Taber, G. Lahav, Threat, anxiety, and support of antiterrorism policies. Am. J. Polit. Sci. **49**(3), 593–608 (2005)
37. E. Bueno de Mesquita, E.S. Dickson, The propaganda of the deed: Terrorism, counterterrorismand mobilization. Am. J. Polit. Sci. **51**(2), 364–381 (2007)

38. K. Gaibulloev, T. Sandler, Hostage taking: determinants of terrorist logistical and negotiation success. J. Peace Res. **46**(6), 739–756 (2009)
39. M. Hughes, Terror and negotiation. Terror. Polit. Violence **2**(1), 72–825 (1990)
40. V.A. Kremenyuk, G.-O. Faure, *International Negotiation: Analysis, Approaches, Issues* (Jossey-Bass San Francisco, California, 2002)
41. E. Pronin, K. Kennedy, S. Butsch, Bombing versus negotiating: how preferences for combating terrorism are affected by perceived terrorist rationality. Basic Appl. Soc. Psychol. **28**(4), 385–392 (2006)
42. D.G. Pruitt, Negotiation with terrorists. Int. Negot. **11**(2), 371–394 (2006)
43. B.I. Spector, Negotiating with villains revisited: research note. Int. Negot. **8**(3), 613–621 (2003)
44. I.W. Zartman, Negotiating with terrorists. Int. Negot. **8**(3), 443–450 (2003)
45. N.A. Bapat, State bargaining with transnational terrorist groups. Int. Stud. Q. **50**(1), 213–230 (2006)
46. T. Sandler et al., Terrorism & game theory. Simul. Gaming **34**(3), 319–337 (2003)
47. T. Boongoen, Q. Shen, C. Price, Disclosing false identity through hybrid link analysis. Artif. Intell. Law **18**(1), 77–102 (2010)
48. T. Boongoen, Q. Shen, Nearest-neighbor guided evaluation of data reliability and its applications. IEEE Trans. Syst. Man Cybern. **40**(6), 1622–1633 (2010)
49. X. Fu, T. Boongoen, Q. Shen, Evidence directed generation of plausible crime scenarios with identity resolution. Appl. Artif. Intell. **24**(4), 253–276 (2010)
50. R.J. Brym, B. Araj, Are suicide bombers suicidal? Stud. Confl. Terror. **35**(6), 432–443 (2012)
51. A. Merari, I. Diamant, A. Bibi, Y. Broshi, G. Zakin, Personality characteristics of self martyrs/suicide bombers and organizers of suicide attacks. Terror. Polit. Violence **22**(1), 87–101 (2009)
52. L. Miller, The terrorist mind II. typologies, psychopathologies, and practical guidelines for investigation. Int. J. Offender Ther. Comp. Criminol. **50**(3), 255–268 (2006)
53. A. Lankford, Could suicide terrorists actually be suicidal? Stud. Confl. Terror. **34**(4), 337–366 (2011)
54. H.V. Savitch, An anatomy of urban terror: lessons from jerusalem and elsewhere. Urb. Stud. **42**(3), 361–395 (2005)
55. H.A. Hagras, A hierarchical type-2 fuzzy logic control architecture for autonomous mobile robots. IEEE Trans. Fuzzy Syst. **12**(4), 524–539 (2004)
56. Z.-Q. Liu, R. Satur, Contextual fuzzy cognitive map for decision support in geographic information systems. IEEE Trans. Fuzzy Syst. **7**(5), 495–507 (1999)
57. R. Diao, Q. Shen, Feature selection with harmony search. IEEE Trans. B Syst. Man Cybern. **42**(6), 1509–1523 (2012)
58. R. Diao, Q. Shen, New approaches to fuzzy-rough feature selection. IEEE Trans. Fuzzy Syst. **17**(4), 824–838 (2009)

Chapter 8
Conclusion

This chapter presents a high-level summary of the work documented in this book. The main contributions include: an innovative concept and approaches for backward fuzzy rule interpolation (BFRI) and its application. This is followed by a discussion about important topics for future work which would help to further improve and expand the current work.

8.1 Summary

This book has presented backward fuzzy rule interpolation, a novel approach that complements traditional FRI by supporting backward inference, allowing flexible interpolation when certain antecedents are missing from the observation. The work is based upon the mechanisms of T-FRI and the α-cut-based FRI method, respectively. The proposed BFRI techniques offer a flexible way in which to perform practical hierarchical fuzzy interpolation. Three detailed BFRI methods were proposed: T-FRI-based BFRI with single missing antecedent value (S-BFRI), T-FRI-based BFRI with multiple missing antecedent values (M-BFRI) and α-cut-based backward fuzzy rule interpolation. The significant problem domain of terrorism risk assessment was chosen as an application area to demonstrate such developments. Publications arising from the work developed in this book are listed in Appendix A.

8.1.1 Transformation Based S-BFRI

In practical applications with interconnected subsets of rules, situations may arise when a crucial antecedent of observation is absent, either due to human error or difficulty in obtaining data, while the associated conclusion may be derived according to alternative rules or even observed directly. If such missing antecedents were involved in the subsequent interpolation process, the final conclusion would not be

S. Jin et al., *Backward Fuzzy Rule Interpolation*,
https://doi.org/10.1007/978-981-13-1654-8_8

deduced using conventional means. However, missing antecedents may be related to certain conclusion and therefore may be inferred or interpolated using the known antecedents and conclusion.

The conceptual definition of such BFRI methods has been proposed for the first time, in this book. The proposed techniques build upon the scale and move transformation-based fuzzy interpolation mechanism. In particular, the approach supports both interpolation and extrapolation which involve multiple intertwined fuzzy rules, with each with multiple antecedents. BFRI with single missing antecedent value (S-BFRI) is proposed for interpolation in situations where the consequent value is known and the values of all but one antecedent variable are also given. S-BFRIcan support intertwined multiple multiantecedent rules involving trapezoidal membership functions. The proposed method supports flexible interpolation when a certain antecedent is missing from the observation, where traditional approaches fail. Four worked examples have been provided to illustrate the operation of this approach.

8.1.2 Transformation Based M-BFRI

In order to handle multiple multiantecedent rules and to ensure the maintenance of convexity and normality of the interpolated outcomes, specific algorithms have been developed to tackle problem scenarios with multiple missing values. Two approaches have been proposed with worked examples provided to illustrate their operations. (1) The parametric approach, which directly extends the S-BFRI method but involves a higher computational complexity; (2) the feedback approach, which is a more generalised method that works more closely with conventional FRI procedures.

Systematic evaluation has been proposed and the consistency, accuracy and robustness of the interpolative procedure are verified by comparing the outcome of the interpolation to the ground truth. This reflects an underlying principle similar to that behind cross-validation and statistical evaluation. Evaluation results show that P-BFRI is more accurate, despite its limited scalability for larger problems. F-BFRI can successfully handle more complex scenarios, and significantly reduces the need for parameter calculation, but its interpolative accuracy is relatively low. The proposed M-BFRI approaches provide more feasible and general means to solve the practical problems associated with multiple missing antecedent values, or multiple antecedent values which need to be inferred or proven.

8.1.3 An Alternative BFRI Method

Although promising the above backward fuzzy rule interpolation concept has only been implemented using an analogical interpolation method. Considering the general versatility and potential of BFRI, it is imperative to extend BFRI to also support the α-cut-based methods. In particular, the fuzzy interpolation technique for multidimensional input spaces (IMUL) is employed to extend the existing BFRI approach. This is done because IMUL allows interpolation using the rules involving multidimensional input spaces. Also, it guarantees that the interpolative outcomes are crisp if the inputs (observations) are also crisp.

A systematic comparison has been carried out between the proposed IMUL-BFRI method and the original implementation using scale and move transformation-based interpolation method (T-FRI-BFRI). The results show that the T-FRI based approach entails slightly higher accuracy than the IMUL-based approach, in terms of both forward and backward methods. Nevertheless, both methods seem to have an acceptable accuracy. Importantly, this investigation has demonstrated that the general concept of BFRI can be potentially achieved via interpolation methods of a different type.

8.1.4 Refinement of Rule Base Based on HBFRI

HBFRI enables unknown antecedent values to be interpolated, given other antecedents and the conclusion. This integrated approach, of hierarchical reasoning and bidirectional interpolative inference, provides a flexible and systematic way of dealing with insufficient information or knowledge that may often appear in real-world problems. The system implementing the proposed technique is able to draw a final conclusion through the exploitation of BFRI even when it is presented with partial observations. A particular concern is due to the high incompleteness of rule bases and the imprecision of observations, as in such situations rules used for interpolation may become inconsistent. A novel approach based on hierarchical bidirectional fuzzy interpolation is therefore proposed in an effort to remove any inconsistencies in a rule base. This also helps identify hidden variables that may be useful during any subsequent intelligent decision support processes. The initial experimental results reported have demonstrated that the proposed method can retain model accuracy while significantly reducing the number of the rules required in the system model.

8.1.5 Application: Terrorism Risk Assessment

BFRI methods have been applied to a realist problem, to build a linguistic evaluation mechanism that can offer a reasonable and adjustable means for terrorist risk assessment. A methodology for qualitatively assessing the risks of terrorism has been proposed. The goal of the study is to predict (and hence, prevent) potential attacks and to provide decision support to the relevant state bodies regarding possible terrorist attack threats. An approach termed hierarchical bidirectional fuzzy rule interpolation (HBFRI) has been described to enable hypothetical reasoning in high-dimensional and sparse rule-based systems. HBFRI is particularly effective in situations where a multiple multiantecedent rule-based system needs to be reconstructed into a multi-layered fuzzy system, and where each of the individual layers is, in general, a sparse rule-based system. HBFRI offers significant flexibility of interpolation where certain crucial antecedents are absent (or assumed to be missing) from given observations. Structured assessments and a realistic case study have been adopted in order to demonstrate the effectiveness of the developed work when applied to terrorist attack prediction or decision support scenarios.

8.2 Future Work

8.2.1 Generalisation of BFRI

It would be useful to develop a generalised approach that can be implemented using other type of interpolation method (e.g. GM [1], FIVE [2], or IRCT [3]), and to compare the results. The proposed method based on IMUL currently can only deal with S-BFRI. An IMUL-based BFRI approach with multiple missing values could be considered in the context of its accuracy and computational complexity compared against those that based on T-FRI.

While in principle, the idea of backward fuzzy rule interpolation (or backward reasoning in general) appears to be applicable to both Mamdani and TSK fuzzy systems [4–7], for the present implementation, the technique described relies on the scale and move transformation-based procedures and is therefore only applicable to Mamdani models. It would be of natural appeal in the light of this to develop the proposed technique for application to TSK fuzzy models.

8.2.2 BFRI Versus Fuzzy Inversion

The underlying concept which supports the proposed BFRI seems to bear close relation to that of fuzzy inversion [8–10]. Inverse fuzzy models can be utilised either as a direct compensation of certain measurement of nonlinearity, or as a controller mechanism for nonlinear plant. Considering FRI as a way of defining “fuzzy function” and the fuzzy rules as “node points” (fuzzy point) of the interpolating function (FRI), the concept of BFRI turns back to the main question of fuzzy function inversion. BFRI discussed in this book focuses on the application domain of interpolation, and could be utilised in conjunction with fuzzy inversion in order to better deal with such problems. A systematic comparative study between these two approaches could offer a significance to the development of backward fuzzy reasoning methodologies.

8.2.3 Antecedent And/Or Rule Selection in Interpolation

Many real-world applications generally involve rules with a large number of antecedents. The error accumulated throughout the interpolation process may affect the accuracy of the final estimation. More importantly, a rule base may consist of less relevant, redundant or even misleading variables, which may further deviate the outcome of an interpolation. Therefore, an intelligent antecedent and/or rule selection procedure may be developed by identifying the most relevant information [11–13], so that the appropriate terms or rules can be determined in order to minimise the overall system complexity.

An initial version of such a new FRI approach [14] has already been presented in order to evaluate the importance of antecedent variables by exploiting FS techniques. A weighted aggregation-based interpolation method is proposed that makes use of the identified antecedent significance values. The original rule base may also be simplified by removing the irrelevant or noisy antecedents using a feature subset search algorithm such as HSFS [12] and retains an antecedent subset of a much lower dimensionality. The present antecedent selection approach for FRI may be further improved by considering unsupervised or semi-supervised FS methods [15–17], which have emerged recently, for analysing the interdependencies between features without the aid of class label information. Fuzzy aggregation functions [11, 18, 19] are of particular assistance in realising this task.

8.2.4 Enhancement of M-BFRI

For BFRI with multiple missing antecedent values, the current techniques are implemented using exhaustive search-based methods, which may be better formulated using advanced solution techniques (e.g. Waltz algorithm [20]) for flexible constraint satisfaction [21]. It may also be further improved via the use of heuristic optimisation algorithms [12, 22–24] that do not require domain discretisation. This may help to obtain even better performance, while reducing the search cost (for P-BFRI in particular). The present work would also benefit from a mechanism for automatic identification and selection of better reasoning paths in handling hierarchical rule models, so that the interpolation process may proceed dynamically [25] according to the current state of the system.

8.2.5 Improve Hierarchical Interpolation Using BFRI

As described in Sect. 1.2 and Chap. 7, hierarchical fuzzy (including interpolation) rule-based systems have reduced computational complexity compared to that of dense fuzzy rule-based systems [26, 27]. However, calculating and representing the intermediate output variable of each layer is still a problem that practical HFS needs to address [22, 24, 28–32]. That is, for many applications, the intermediate variables usually do not possess any actual meaning and consequently make HFS difficult to design. These intermediate variables without "actual meaning" also reduce the transparency of fuzzy systems, and HFS becomes less understandable and interpretable. It may be beneficial to investigate how the proposed BFRI based reverse reasoning [33] might be used to form an alternative theoretical basis, upon which a hierarchical fuzzy interpolation mechanism would be developed. This would help resolve this "intermediate variable values" problem.

8.2.6 BFRI Based Rule Base Refinement

Inconsistent rules have similar conditions but different consequences. It is essential for learning mechanisms to identify possible conflicts in rule bases and to obtain good logical coherence. For this purpose, a numerical assessment termed inconsistency index has been introduced [34], which helps to establish the consistency or inconsistency of rule bases. Inference engines and fuzzy rule bases are two key parts, either in a conventional fuzzy system or in a fuzzy interpolation system. Reasoning inconsistency may exist in an inference engine, and the work in [35] proposed a method to modify the identified interpolated rules in an effort to restore reasoning consistency.

However, the above method relies upon the assumption that all rules for interpolation (which are provided in the initial rule base) are true and fixed. This may not always be the case. A new rule which is inferred or interpolated from the original rule base may conflict with an existing rule. A range of approaches for rule pruning by applying reverse reasoning techniques have been proposed [36–38]. From this observation, it would be useful to employ BFRI to refine a fuzzy rule base, inferring the values of the existing antecedents according to the inconsistent conclusion. Then, the biases (or regressive bias) between the obtained values and the ground truth can be applied to refine the desired consequence.

8.2.7 "Many-to-One" Problem in BFRI

The underlying problem that BFRI attempts to address is fundamentally that of "many-to-one", where a number of different value combinations, for the antecedent variables, may lead to very similar observed values for the consequent variable. Although this issue has been partly analysed from an experimental viewpoint in this book, much remains to be done. In particular, the problem may be exacerbated, if the number of missing values becomes large. This makes it very challenging to restore the "true" original observation. Fortunately, different observations obtained via the BFRI process will generate the same or very similar outcomes. Thus, they may be regarded as "equally possible" given the limited amount of knowledge with regard to an application problem at hand. Nevertheless, it is important to be able to improve the proposed approach in an effort to better handle the "many-to-one" problem. This will be of practical significance for BFRI when employed for accuracy-critical applications (e.g. medical diagnosis [39]).

8.2.8 Further Applications of FRI/BFRI

The prediction of terrorist attack and decision support in counter-terrorism, as described in Chap. 7, is just one application instance of backward fuzzy interpolation. For instance, it can be very challenging to resolve the puzzle of a given crime from a set of dynamic evidence and finally make a reasonable decision. A rule-based system, such as the one used in the current application on terrorism risk assessment, may not be able to address certain unprecedented scenarios or effectively respond to sudden changes of political attitude. For such cases, dynamic interpolation [25] technique may be adopted in order to enhance the reasoning system. Intuitively, to further evaluate the potential of this research, it may be interesting to apply the technique to many other real-world application domains also, such as medical diagnosis [39], network intrusion detection [40] and oil exploration [41].

References

1. P. Baranyi, L.T. Kóczy, T.D. Gedeon, A generalized concept for fuzzy rule interpolation. IEEE Trans. Fuzzy Syst. **12**(6), 820–837 (2004)
2. S. Kovács, Extending the fuzzy rule interpolation "five" by fuzzy observation. Comput. Intell. Theory Appl. **38**, 485–497 (2006)
3. S. Chen, Y. Ko, Fuzzy interpolative reasoning for sparse fuzzy rule-based systems based on α-cuts and transformations techniques. IEEE Trans. Fuzzy Syst. **16**(6), 1626–1648 (2008)
4. G. Feng, A survey on analysis and design of model-based fuzzy control systems. IEEE Trans. Fuzzy Syst. **14**(5), 676–697 (2006)
5. F. Hoffmann, D. Schauten, S. Holemann, Incremental evolutionary design of tsk fuzzy controllers. IEEE Trans. Fuzzy Syst. **15**(4), 563–577 (2007)
6. Y. Jin, Fuzzy modeling of high-dimensional systems: complexity reduction and interpretability improvement. IEEE Trans. Fuzzy Syst. **8**(2), 212–221 (2000)
7. T.A. Johansen, R. Shorten, R. Murray-Smith, On the interpretation and identification of dynamic takagi-sugeno fuzzy models. IEEE Trans. Fuzzy Syst. **8**(3), 297–313 (2000)
8. P. Baranyi, P. Korondi, H. Hashimoto, M. Wada, Fuzzy inversion and rule base reduction, in *Proceedings of International Conference on Intelligent Engineering Systems* (1997), pp. 301–306
9. A.R. Várkonyi-Kóczy, A. Almos, T. Kovácsházy, Genetic algorithms in fuzzy model inversion, in *Proceedings of International Conference on Fuzzy Systems*, vol. 3 (1999), pp. 1421–1426
10. A.R. Várkonyi-Kóczy, G. Péceli, T.P. Dobrowiecki, T. Kovácsházy, Iterative fuzzy model inversion, in *Proceedings of International Conference on Fuzzy Systems*, vol. 1 (1998), pp. 561–566
11. T. Boongoen, Q. Shen, Nearest-neighbor guided evaluation of data reliability and its applications. IEEE Trans. Syst. Man Cybern. **40**(6), 1622–1633 (2010)

12. R. Diao, Q. Shen, Feature selection with harmony search. IEEE Trans. Syst. Man Cybern. B **42**(6), 1509–1523 (2012)
13. N.M. Parthalain, R. Jensen, Simultaneous feature and instance selection using fuzzy-rough bireducts, in *Proceedings of International Conference on Fuzzy Systems* (2013), pp. 1–7
14. R. Diao, S. Jin, Q. Shen, Antecedent selection in fuzzy rule interpolation using feature selection techniques, in *Proceedings of IEEE International Conference on Fuzzy Systems* (2014), pp. 2206–2213
15. N. Mac Parthaláin, R. Jensen, Measures for unsupervised fuzzy-rough feature selection. Int. J. Hybrid Intell. Syst. **7**(4), 249–259 (2010)
16. N. Mac Parthaláin, R. Jensen, Fuzzy-rough set based semi-supervised learning, in *IEEE International Conference on Fuzzy Systems* (2011), pp. 2465–2472
17. J. Zhao, K. Lu, X. He, Locality sensitive semi-supervised feature selection. Neurocomputing **71**(10–12), 1842–1849 (2008)
18. Y. Narukawa, in *Modeling Decisions: Information Fusion and Aggregation Operators*. Cognitive Technologies (Springer, 2010)
19. R. Yager, On ordered weighted averaging aggregation operators in multicriteria decisionmaking. IEEE Trans. Syst. Man Cybern. **18**(1), 183–190 (1988)
20. D. Waltz, Understanding line drawings of scenes with shadows, in *The Psychology of Computer Vision* (McGraw-Hill, 1975), pp. 11–91
21. I. Miguel, Q. Shen, Fuzzy rrDFCSP and planning. Artif. Intell. **148**(1), 11–52 (2003)
22. M. Lee, H. Chung, F. Yu, Modeling of hierarchical fuzzy systems. Fuzzy Sets Syst. **138**(2), 343–361 (2003)
23. M. Wagenknecht, K. Hartmann, Fuzzy modelling with tolerances. Fuzzy Sets and Syst. **20**(3), 325–332 (1986)
24. D. Wang, X. Zeng, J. Keane, Intermediate variable normalization for gradient descent learning for hierarchical fuzzy system. IEEE Trans. Fuzzy Syst. **17**(2), 468–476 (2009)
25. N. Naik, R. Diao, C. Quek, Q. Shen, Towards dynamic fuzzy rule interpolation, in *Proceedings of International Conference on Fuzzy Systems* (2013), pp. 1–7
26. L. Wang, Universal approximation by hierarchical fuzzy systems. Fuzzy Sets Syst. **93**(2), 223–230 (1998)
27. L. Wang, Analysis and design of hierarchical fuzzy systems. IEEE Trans. Fuzzy Syst. **7**(5), 617–624 (1999)
28. M.G. Joo, A method of converting conventional fuzzy logic system to 2 layered hierarchical fuzzy system, in *Proceedings of International Conference on Fuzzy Systems*, vol. 2 (2003), pp. 1357–1362
29. D. Wang, X.-J. Zeng, J.A. Keane, Learning for hierarchical fuzzy systems based on the gradient-descent method, in *Proceedings of International Conference on Fuzzy Systems* (2006), pp. 92–99
30. Y.J.W.W.H. Wang, S. Kwong, K.F. Man, Multi-objective hierarchical genetic algorithm for interpretable fuzzy rule-based knowledge extraction. Fuzzy Sets Syst. **149**(1), 149–186 (2005)
31. X.-J. Zeng, J.Y. Goulermas, P. Liatsis, D. Wang, J.A. Keane, Hierarchical fuzzy systems for function approximation on discrete input spaces with application. IEEE Trans. Fuzzy Syst. **16**, 1197–1215 (2008)
32. X.-J. Zeng, J.A. Keane, Approximation capabilities of hierarchical fuzzy systems. IEEE Trans. Fuzzy Syst. **13**, 659–672 (2005)
33. T. Arnould, S. Tano, Interval-valued fuzzy backward reasoning. IEEE Trans. Fuzzy Syst. **3**(4), 425–437 (1995)
34. N. Xiong, L. Litz, Reduction of fuzzy control rules by means of premise learning-method and case study. Fuzzy Sets Syst. **132**(2), 217–231 (2002)
35. N. Xiong, L. Litz, Adaptive fuzzy interpolation. IEEE Trans. Fuzzy Syst. **19**(6), 1107–1126 (2011)
36. S. Chawla, J.G. Davis, G. Pandey, On local pruning of association rules using directed hypergraphs, in *ICDE*, vol. 4 (2004), pp. 832–841

37. A. Di Nola, W. Pedrycz, S. Sessa, Fuzzy relation equations with equality and difference composition operators. Fuzzy Sets Syst. **25**(2), 205–215 (1988)
38. L. Fu, Rule generation from neural networks. IEEE Trans. Syst. Man Cybern. **24**(8), 1114–1124 (1994)
39. I. Gadaras, L. Mikhailov, An interpretable fuzzy rule-based classification methodology for medical diagnosis. Artif. Intell. Med. **47**(1), 25–41 (2009)
40. A. Tajbakhsh, M. Rahmati, A. Mirzaei, Intrusion detection using fuzzy association rules. Appl. Soft Comput. **9**(2), 462–469 (2009)
41. K.W. Wong, D. Tikk, T.D. Gedeon, L.T. Kóczy, Fuzzy rule interpolation for multidimensional input spaces with applications: a case study. IEEE Trans. Fuzzy Syst. **13**(6), 809–819 (2005)

Appendix A
Glossary of Terms

α-cut
The α-cut A_α ($\alpha \in (0, 1]$) of a fuzzy set A is a crisp set; it contains the elements of the universe of discourse with membership degree not smaller than α; formally, it is defined as: $A_\alpha = \{x|A(x) \geq \alpha, \alpha \in (0, 1]\}$.

AI (Artificial Intelligence)
Artificial intelligence [1, 2] is a branch of computer science which is concerned with intelligent systems and the development of theories, technologies and applications for complex problems. Research in the area of AI extends to domains for reasoning, knowledge discovery, robotics, perception, natural language processing, expert system, etc.

Backward Reasoning
Backward reasoning [3] is goal-driven, as the aim is to deduce a particular goal using rules. The advantage of backward reasoning is that only information relative to the goal to prove is considered. To prove a particular goal, backward reasoning is therefore more efficient than forward reasoning, due to the lower amount of information that it is necessary to consider.

EP (Extension Principle)
Extension Principle [4, 5] extends a conventional mapping function $f : A_1 \times \cdots \times A_n \rightarrow B$, where $A_1, \ldots, A_n$ and B are crisp domains, to a fuzzy mapping f^*: $\mu_{A_1}(x_1) \times \cdots \times \mu_{A_n}(x_n) \rightarrow \mu_B(y)$ where, for every element x_i, $\mu_{A_i}(x_i)$ denotes the membership of x_i with respect to (fuzzy) set A_i. Formally, the formula to define f^* is: $\mu_B(y) = sup_{x_1,\ldots,x_n \rightarrow y} min\{\mu_{A_1}(x_1), \ldots, \mu_{A_n}(x_n)\}$, for all $x_1 \in A_1, \ldots, x_n \in A_n$ and $y \in B$.

Forward Reasoning
Forward reasoning [3] is used to prove a particular goal and is data-driven. To prove a goal, it is first necessary to deduce all the information that it is possible to infer, using all the available rules and data. The second step consists of checking whether

S. Jin et al., *Backward Fuzzy Rule Interpolation*,
https://doi.org/10.1007/978-981-13-1654-8

the goal to prove is among the newly deduced information. However, much useless information is also deduced, leading to a poor performance of the method.

FS (Feature Selection)

Feature selection [6] is a commonly used approach in machine learning (may also be known as feature subset selection, variable selection or attribute reduction) and can be considered as the process of selecting the input attributes of a data set that most closely define a particular outcome.

Fuzzy Set Theory

Fuzzy sets [4] are sets whose elements have degrees of membership. Fuzzy set theory was introduced by Lotfi A. Zadeh in 1965 as an extension of the classical set. In traditional set theory, the membership of elements in a set is defined in binary terms according to a hard condition an element either belongs to the set or an element does not belong to the set. In contrast, fuzzy set theory allows gradual membership of elements in a set; this is described by employing a membership function in the real unit interval [0, 1].

RP (Resolution Principle)

Resolution Principle [7, 8] is a method of theorem proving that proceeds by constructing refutation proofs. Resolution principle applies to first-order logic formulas in Skolemised form. These formulas are basically sets of clauses each of which is a disjunction of literals. Unification is a key technique in proofs by resolution.

In KH fuzzy rule interpolation method, RP describes the decomposition of fuzzy sets to α-cuts:

$$A = \bigcup_{\alpha \in [0,1]} \alpha A_\alpha \tag{A.1}$$

where $\bigcup$ means maximum.

SBR (Similarity-Based Reasoning)

Similarity-based reasoning [9, 10] aims at studying which kinds of logical consequence relations make sense when taking into account that some propositions may be closer to be true than others. Essentially, SBR has been investigated from two different perspectives: qualitative approaches and quantitative approaches.

t-norm

t-norm [11] as the abbreviation of triangular norm is a kind of binary operation used in the framework of probabilistic metric spaces and in multivalued logic, specifically in fuzzy logic. A t-norm generalises intersection in a lattice and conjunction in logic. The name triangular norm refers to the fact that in the framework of probabilistic metric spaces t-norms are used to generalise triangle inequality of ordinary metric spaces. t-norm is a binary operation T on the interval [0, 1] satisfying the following conditions:

$$\begin{aligned} T(x, y) &= T(y, x) \quad (commutativity) \\ T(x, T(y, z)) &= T(T(x, y), z) \quad (associativity) \\ T(x, y) \leq T(x, z) \quad whenever \quad y &\leq z \quad (monotonicity) \\ T(x, 1) &= x \quad (neutral \ \ element \ \ 1) \end{aligned} \tag{A.2}$$

Appendix B
Examples for Calculations

1. **Example of calculations for Table** 3.3

Here, the value $\omega_{A_1^1} = 0.27$ is used as an example to illustrate the calculation process.

1. According to Eq. 2.75: $d(A_k^i, A_k^*) = d(Rep(A_k^i), Rep(A_k^*))$, calculate the distance between A_1^* and A_1^1: $d(A_1^1, A_1^*)$. In terms of the values in Table. 3.2 and Eq. 2.56, $d(A_1^1, A_1^*) = \left| \frac{3.5+\frac{4.0+5.0}{2}+7.0}{3} - \frac{0.2+\frac{1.1+2.2}{2}+2.7}{3} \right| = 3.483$.
2. Calculate the term weight $\omega'_{A_1^1}$ by using Eq. 2.75: $\omega'_{A_k^i} = 1/d(A_k^i, A_k^*)$. Therefore, $\omega'_{A_1^1} = 1/d(A_1^1, A_1^*) = 1/3.483 = 0.287$.
3. Calculate the sum of the term weights of the kth dimension $\sum_{i=1}^{N} \omega'_{A_k^i}$. Here, for the first antecedent dimension, $\sum_{i=1}^{4} \omega'_{A_1^i} = \omega'_{A_1^1} + \omega'_{A_1^2} + \omega'_{A_1^3} + \omega'_{A_1^4} = 0.287 + 0.417 + 0.211 + 0.146 = 1.061$.
4. Finally, calculate the normalised weight $\omega_{A_1^1}$ by using Eq. 2.74: $\omega_{A_k^i} = \frac{\omega'_{A_k^i}}{\sum_{i=1}^{N} \omega'_{A_k^i}}$.

$$\omega_{A_1^1} = \frac{\omega'_{A_1^1}}{\sum_{i=1}^{4} \omega'_{A_1^i}} = 0.287/1.061 = 0.270.$$

2. **Calculations for Example** 5.1.1

The backward interpolation process in Example 5.1.1 is detailed below.

Table 3.3 Normalised weights for the given antecedents

	R_1	R_2	R_3	R_4
B	0.17	0.45	0.23	0.15
A_1	0.27	0.39	0.20	0.14
A_2	0.23	0.35	0.26	0.16
A_4	0.15	0.36	0.40	0.09

S. Jin et al., *Backward Fuzzy Rule Interpolation*,
https://doi.org/10.1007/978-981-13-1654-8

1. According to Eq. 5.18, $\lambda_{\text{core}B} = \dfrac{b_0^* - b_0^1}{b_0^2 - b_0^1} = \dfrac{\frac{6.5+7.0}{2} - \frac{2.0+2.5}{2}}{\frac{13+13.5}{2} - \frac{2.0+2.5}{2}} = 0.409.$
2. According to Eq. 5.19, the reference point $a_{3,0}^*$:
   $a_{3,0}^* = a_{3,0}^1 + \sqrt{\left|4(\lambda_{\text{core}B})^2(b_0^* - b_0^1)^2 - \sum_{i=1,i\neq3}^{4}(a_{i,0}^* - a_{i,0}^1)^2\right|}$
   $= \frac{1.5+2.0}{2} + \sqrt{|4 \times 0.409^2 \times (6.75 - 2.25)^2 - (4.5 - 1.6)^2 - (5.75 - 2.25)^2 - (5.85 - 1.8)^2|}$
   $= 7.56.$
3. The right core $a_{3,1}^*$ can be derived from Eqs. 5.20 and 5.21.
   First, $\lambda_{\text{right}B} = \dfrac{b_1^* - b_1^1}{b_1^2 - b_1^1} = \dfrac{7.0 - 2.5}{13.5 - 2.5} = 0.440$
   Then, $a_{3,1}^* = a_{3,1}^1 + \sqrt{\left|4 \times (\lambda_{\text{right}B})^2 \times (b_1^* - b_1^1)^2 - \sum_{i=1,i\neq3}^{4}(a_{i,1}^* - a_{i,1}^1)^2\right|}$
   $= 2.0 + \sqrt{|4 \times (0.440)^2 \times (7.0 - 2.5)^2 - (5.0 - 2.2)^2 - (6.0 - 2.5)^2 - (6.5 - 2.1)^2|}$
   $= 8.04$
4. The left core $a_{3,-1}^*$ can then be derived from Eqs. 5.22 and 5.23.
   $a_{3,-1}^* = a_{3,-1}^1 + \sqrt{\left|4(\lambda_{\text{left}B})^2(b_{-1}^* - b_{-1}^1)^2 - \sum_{i=1,i\neq3}^{4}(a_{i,-1}^* - a_{i,-1}^3)^2\right|}$
   $= 1.5 + \sqrt{\left|4 \times (\frac{6.5-2.0}{13.0-2.0})^2(6.5 - 2.0)^2 - (4.0 - 1.1)^2 - (5.5 - 2.0)^2 - (5.2 - 1.5)^2\right|}$
   $= 7.37$
5. According to Eqs. 5.7, 5.25 and 5.26, the right flank $a_{3,2}^*$ is calculated as below:
   $r_{B^*} = b_2^* - b_1^* = 8.7 - 7.0 = 1.70$
   $r_3 = \sqrt{\left|r_{B^*}^2 - \sum_{i=1,i\neq3}^{4}(a_{i,2}^* - a_{i,1}^*)^2\right|}$
   $= \sqrt{|1.7^2 - (7.0 - 5.0)^2 - (7.5 - 6.0)^2 - (7.5 - 6.5)^2|}$
   $= 2.08$
   Then, $a_{3,2}^* = r_3 + a_{3,1}^* = 2.08 + 8.04 = 10.13$
6. According to Eqs. 5.27, the left flank $a_{3,-2}^*$ can be obtained.
   $r'_{B^*} = b_{-1}^* - b_{-2}^* = 6.5 - 5.5 = 1.00$
   $r'_3 = \sqrt{\left|(r'_{B^*})^2 - \sum_{i=1,i\neq3}^{4}(a_{i,-1}^* - a_{i,-2}^*)^2\right|}$
   $= \sqrt{|1.0^2 - (4.0 - 3.5)^2 - (5.5 - 5.0)^2 - (5.2 - 4.5)^2|}$
   $= 0.49$
   Then, $a_{3,-2}^* = a_{3,-1}^* - r'_3 = 7.37 + 0.49 = 6.88$
7. Finally, the backward interpolative conclusion:
   $A_3^* = (a_{3,-2}^*, a_{3,-1}^*, a_{3,1}^*, a_{3,2}^*) = (6.88, 7.37, 8.04, 10.12).$

Reference

1. M. Hutter, *Universal Artificial Intelligence* (Springer, Berlin, 2005)
2. S.J. Russell, P. Norvig, *Artificial Intelligence: A Modern Approach*, 2nd edn. (Prentice Hall, Upper Saddle River, New Jersey, 2003)
3. T. Arnould, S. Tano, Interval-valued fuzzy backward reasoning. IEEE Trans. Fuzzy Syst. **3**(4), 425–437 (1995)
4. L. Zadeh, Fuzzy sets. Inf. Control **8**(3), 338–353 (1965)
5. L. Zadeh, Fuzzy logic and approximate reasoning. Synthese **30**(3–4), 407–428 (1975)
6. D. Paul, E. Bair, T. Hastie, R. Tibshirani, Preconditioning for feature selection and regression in high-dimensional problems. Ann. Stat. **36**(4), 1595–1618 (2008)
7. J. Robinson, A machine-oriented logic based on the resolution principle. J. ACM (JACM) **12**(1), 23–41 (1965)
8. Z. Shen, L. Ding, M. Mukaidono, Fuzzy resolution principle, in *Proceedings of the Eighteenth International Symposium on Multiple-Valued Logic* (IEEE, 1988), pp. 210–215
9. M.-G. Chun, A similarity-based bidirectional approximate reasoning method for decision-making systems. Fuzzy Sets Syst. **117**(2), 269–278 (2001)
10. S. Raha, N.R. Pal, K.S. Ray, Similarity-based approximate reasoning: methodology and application. IEEE Trans. Syst. Man Cybern. Part A Syst. Hum. **32**(4), 541–547 (2002)
11. E.P. Klement, R. Mesiar, E. Pap, Triangular norms. position paper i: basic analytical and algebraic properties. Fuzzy Sets Syst. **143**(1), 5–26 (2004)

References

Printed in the United States
By Bookmasters